Department for Economic
and Social Information and
Policy Analysis

Statistical Division

World Statistics in Brief   Series V   No. 16

World Statistics Pocketbook

Containing data available
as of 31 December 1994

United Nations · New York 1995

The designations employed and the presentation of material in this publication do not imply the expression of any opinion whatsoever on the part of the Secretariat of the United Nations concerning the legal status of any country, territory, city or area or of its authorities, or concerning the delimitation of its frontiers or boundaries. The term "country" as used in this publication also refers, as appropriate, to territories or areas.
The designations "developed" and "developing" regions are intended for statistical convenience and do not necessarily express a judgement about the stage reached by a particular country or area in the development process.

ST/ESA/STAT/SER.V/16
United Nations Publication
Sales No. E.95.XVII.7 ISBN 92-1-161376-0
Copyright © United Nations, 1995
All rights reserved
Manufactured in the United States of America
Inquiries should be directed to:
United Nations Publications
New York, NY 10017

Preface

This *World Statistics Pocketbook* is the fifteenth compilation of key basic statistical series for countries and regions of the world issued by the United Nations. It responds to General Assembly resolution 2626 (XXV), in which the Secretary-General is requested to supply basic national data that will increase international public awareness of countries' development efforts. The *World Statistics Pocketbook* serves as a vehicle for disseminating, in compact form, important basic economic and social indicators relating to most countries and areas of the world.

The indicators shown are selected from the wealth of international statistical information compiled regularly by the Statistical Division, Department for Economic and Social Information and Policy Analysis of the United Nations Secretariat, and the statistical services of the United Nations specialized agencies and of other international organizations and institutions.

This issue of the *World Statistics Pocketbook* generally covers the years 1985 and 1994. The statistics included for each year shown are those most recently compiled and made available by the international statistical services from official national sources, supplemented by international estimates in some fields. Statistical sources and methods are described in the section "Technical notes" and in footnotes. Statistics presented are in general the latest available to the Statistical Division of the United Nations Secretariat as of 31 December 1994.

Readers wishing to consult more detailed statistics and descriptions of technical methods used in their collection and compilation are referred to the more specialized publications listed in the Introduction and in the section "Statistical sources" at the end of this publication.

It is hoped that the public in general and students will find this dissemination of basic statistics and indicators informative, useful and rewarding.

Contents

Preface v
Tables viii
Introduction xiii
Explanatory notes and abbreviations xiv
Conversion coefficients and factors xiv
Country and area tables 1
Technical notes 207
    Geographical coverage 207
    General indicators 208
    Economic indicators 208
    Social indicators 212
    Environmental indicators 214
Statistical sources 217
Current United Nations statistical publications 219

Tables

Afghanistan	2
Albania	3
Algeria	4
American Samoa	5
Angola	6
Antigua and Barbuda	7
Argentina	8
Armenia	9
Australia	10
Austria	11
Azerbaijan	12
Bahamas	13
Bahrain	14
Bangladesh	15
Barbados	16
Belarus	17
Belgium	18
Belize	19
Benin	20
Bhutan	21
Bolivia	22
Bosnia-Herzegovina	23
Botswana	24
Brazil	25
Brunei Darussalam	26
Bulgaria	27
Burkina Faso	28
Burundi	29
Cambodia	30
Cameroon	31
Canada	32
Cape Verde	33
Central African Republic	34
Chad	35
Chile	36
China	37
Colombia	38
Comoros	39
Congo	40
Cook Islands	41
Costa Rica	42
Côte d'Ivoire	43
Croatia	44
Cuba	45

Cyprus	46
Czech Republic	47
Denmark	48
Djibouti	49
Dominica	50
Dominican Republic	51
East Timor	52
Ecuador	53
Egypt	54
El Salvador	55
Equatorial Guinea	56
Eritrea	57
Estonia	58
Ethiopia	59
Fiji	60
Finland	61
France	62
French Guiana	63
French Polynesia	64
Gabon	65
Gambia	66
Georgia	67
Germany	68
Ghana	69
Greece	70
Grenada	71
Guadeloupe	72
Guam	73
Guatemala	74
Guinea	75
Guinea-Bissau	76
Guyana	77
Haiti	78
Honduras	79
Hong Kong	80
Hungary	81
Iceland	82
India	83
Indonesia	84
Iran (Islamic Republic of)	85
Iraq	86
Ireland	87
Israel	88
Italy	89
Jamaica	90
Japan	91
Jordan	92
Kazakhstan	93

Kenya	94
Kiribati	95
Korea, Democratic People's Republic of	96
Korea, Republic of	97
Kuwait	98
Kyrgyzstan	99
Lao People's Democratic Republic	100
Latvia	101
Lebanon	102
Lesotho	103
Liberia	104
Libyan Arab Jamahiriya	105
Liechtenstein	106
Lithuania	107
Luxembourg	108
Macau	109
Madagascar	110
Malawi	111
Malaysia	112
Maldives	113
Mali	114
Malta	115
Marshall Islands	116
Martinique	117
Mauritania	118
Mauritius	119
Mexico	120
Micronesia, Federated States of	121
Monaco	122
Mongolia	123
Morocco	124
Mozambique	125
Myanmar	126
Namibia	127
Nauru	128
Nepal	129
Netherlands	130
Netherlands Antilles	131
New Caledonia	132
New Zealand	133
Nicaragua	134
Niger	135
Nigeria	136
Northern Mariana Islands	137
Norway	138
Oman	139
Pakistan	140
Palau	141

Panama	142
Papua New Guinea	143
Paraguay	144
Peru	145
Philippines	146
Poland	147
Portugal	148
Puerto Rico	149
Qatar	150
Republic of Moldova	151
Réunion	152
Romania	153
Russian Federation	154
Rwanda	155
Saint Kitts/Nevis	156
Saint Lucia	157
Saint Vincent/Grenadines	158
Samoa	159
San Marino	160
Sao Tome and Principe	161
Saudi Arabia	162
Senegal	163
Seychelles	164
Sierra Leone	165
Singapore	166
Slovakia	167
Slovenia	168
Solomon Islands	169
Somalia	170
South Africa	171
Spain	172
Sri Lanka	173
Sudan	174
Suriname	175
Swaziland	176
Sweden	177
Switzerland	178
Syrian Arab Republic	179
Tajikistan	180
Thailand	181
The former Yugoslav Republic of Macedonia	182
Togo	183
Tonga	184
Trinidad and Tobago	185
Tunisia	186
Turkey	187
Turkmenistan	188
Uganda	189

Ukraine	190
United Arab Emirates	191
United Kingdom	192
United Republic of Tanzania	193
United States of America	194
Uruguay	195
US Virgin Islands	196
Uzbekistan	197
Vanuatu	198
Venezuela	199
Viet Nam	200
Western Sahara	201
Yemen	202
Yugoslavia	203
Zaire	204
Zambia	205
Zimbabwe	206

Introduction

Although considerable progress has been made towards standardization of statistical definitions worldwide, concepts and definitions employed at the national level in the collection of data often differ significantly from country to country. In the *World Statistics Pocketbook* it is not possible to indicate where deviations occur from the standard definitions, for reasons of space and brevity.

Short descriptions of the standard definitions for selected items are provided in the "Technical notes", beginning on page 207.

Readers interested in more detailed data and information regarding their coverage should consult the following major publications:

United Nations
- *Statistical Yearbook*
  (United Nations publication, Series S) [14]*
- *Demographic Yearbook*
  (United Nations publication, Series R) [7]
- *Industrial Statistics Yearbook*
  (United Nations publication, Series P) [9]
- *National Accounts Statistics*
  (United Nations publication, Series X) [13]
- *International Trade Statistics Yearbook*
  (United Nations publication, Series G) [10]
- *Energy Statistics Yearbook*
  (United Nations publication, Series J) [8]
- *Monthly Bulletin of Statistics*
  (United Nations publication, Series Q) [12]
- *World Population Prospects* [15]
- *World Urbanization Prospects* [16]

Food and Agriculture Organization of the United Nations (Rome)
- *FAO yearbook: Production* [1]

International Labour Office (Geneva)
- *Year Book of Labour Statistics* [3]

International Monetary Fund (Washington DC)
- *International Financial Statistics* (monthly) [4]

United Nations Educational, Scientific and Cultural Organization (Paris)
- *Statistical Yearbook* [17]

World Tourism Organization (Madrid)
- *Yearbook of Tourism Statistics* [21]

---

* Numbers in brackets refer to numbered entries listed in "Statistical sources" at the end of this publication.

Explanatory notes and abbreviations

...	Data not available
—	Magnitude zero or negligible or not applicable
0.0	Magnitude not zero, but less than half of the unit employed
Per cent (%) p.a.	Per cent per annum
Km²	Square kilometre
1 000 Mt	Thousand metric tons
Mill Mt	Million metric tons

Decimal figures are always preceded by a period (.).

Conversion coefficients and factors

The metric system of weights and measures has been employed in the *World Statistics Pocketbook*. The following table shows the equivalents of the basic metric, British imperial and United States units of measurement:

Area
　　1 square km. = 0.386 102 square mile
　　　　1 hectare = 2.471 054 acres

Volume
　　1 cubic metre = 35.314 667 cubic feet or
　　　　　　　　 = 1.307 951 cubic yards

Weight or mass
　　　　　1 ton = 1.102 311 short tons, or
　　　　　　　 = 0.987 207 long ton
　　　　1 gram = 0.032 151 troy ounce
　　1 kilogram = 35.273 962 avdp. ounces
　　　　　　　 = 2.204 623 avdp. pounds

Distance
　　1 kilometre = 0.621 371 mile

Country and area tables

Afghanistan

Region Southern Asia
Location (longitude, latitude) 34°53'N 69°17'E
Currency afghani
Population (1995 est., in thousands) 23196
Surface area (square kms) 652090
Population density (pop. per square km) 25
Sex ratio (females per 100 males) 95
Largest city (pop. in thousands) Kabul (1565)
UN membership date 19-Nov-1946
Major language(s) Pashto, Persian, Uzbek

Economic indicators	1985	1994
Exchange rate (US$)	50.60	50.60
Consumer price index (1980=100)	219	818[a]
Balance of payments, current account (million US$)	−243	...
Tourist arrivals (in thousands)	9[b]	6[c]
Economically active female population (%)	7.0[d]	9.2
Economically active male population (%)	87.7[d]	86.0
Annual growth of econ. active female pop. (%)	3.7[a]	...
Annual growth of econ. active male pop. (%)	2.1[a]	...
Agricultural production index (1979-1981=100)	77[b]	75[c]
Food production index (1979-1981=100)	78[b]	75[c]
Commercial energy production (1,000 Mt coal equiv)	3710[d]	454[e]
Motor vehicles (per 1,000 population)	4	3[a]
Telephone lines (per 100 inhabitants)	0.2[f]	0.2[e]

Social indicators	1990/95
Growth rate of population (% per annum)	6.7
Age group 0-14 years (%)	40.0
Age group 60+ years (women and men, %)	2.3/2.2
Life expectancy at birth (women and men, years)	44/43
Infant mortality rate (per 1,000 births)	162
Total fertility rate (births per woman)	7
Urban population (%)	20
Urban population growth rate (% per annum)	8.6
Rural population growth rate (% per annum)	6.3
Foreign-born (1985,%)	0.2
Refugees	60000
Primary-secondary gross enrolment ratio (f and m, per 100)	13/26[b]
Third-level students (per 100,000 population)	147
Newspaper circulation (per 1,000 population)	11
Television receivers (per 1,000 population)	8

Environment	1990/91
Threatened species	31
Forested area (%)	2.9
CO_2 emissions (10,000 Mt)	1405
Energy consumption (1,000 Mt coal equiv)	72
Precipitation (mm)	372[f]
Average temperature (January and July, centigrade)	−2.3/24.8[d]

a 1990. b 1988. c 1992. d 1980. e 1991. f 1987.

Albania

Region	Southern Europe
Location (longitude, latitude)	41°34'N 19°84'E[a]
Currency	lek
Population (1995 est., in thousands)	3390
Surface area (square kms)	28748
Population density (pop. per square km)	115
Sex ratio (females per 100 males)	95
Largest city (pop. in thousands)	Tirana (245)[a]
UN membership date	14-Dec-1955
Major language(s)	Albanian

Economic indicators	1985	1994
Exchange rate (US$)	7.00	84.00
Consumer price index (1980=100)	100[b]	442
Balance of payments, current account (million US$)	−36	−32[c]
GDP (million US$)	2408	810[d]
GDP (per capita US$)	813	247[d]
Long-term rate of change in GDP (% per annum)	1.8	−21.7[d]
Gross fixed capital formation (% of GDP)	33.2	34.6[d]
Economically active female population (%)	55.6[e]	59.0
Economically active male population (%)	82.9[e]	84.9
Annual growth of econ. active female pop. (%)	3.1[b]	...
Annual growth of econ. active male pop. (%)	2.8[b]	...
Labour force in industry (%)	45.6[e]	43.3[d]
Labour force in agriculture (%)	20.9[e]	23.0[d]
Agricultural production index (1979-1981=100)	102[f]	98[c]
Food production index (1979-1981=100)	104[f]	102[c]
Commercial energy production (1,000 Mt coal equiv)	4722[e]	2830[d]
Motor vehicles (per 1,000 population)	1	1[b]
Telephone lines (per 100 inhabitants)	1.2[g]	1.3[d]

Social indicators	1990/95
Growth rate of population (% per annum)	0.8
Age group 0-14 years (%)	31.2
Age group 60+ years (women and men, %)	4.6/3.9
Life expectancy at birth (women and men, years)	77/71
Infant mortality rate (per 1,000 births)	23
Total fertility rate (births per woman)	3
Urban population (%)	37
Urban population growth rate (% per annum)	1.7
Rural population growth rate (% per annum)	0.3
Foreign-born (1985,%)	0.4
Refugees	3000
Primary-secondary gross enrolment ratio (f and m, per 100)	90/94
Third-level students (per 100,000 population)	680
Newspaper circulation (per 1,000 population)	42
Television receivers (per 1,000 population)	87
Parliamentary seats (women and men, %)	6/94

Environment	1990/91
Threatened species	25
Forested area (%)	36.4
CO_2 emissions (10,000 Mt)	1705
Energy consumption (1,000 Mt coal equiv)	872
Precipitation (mm)	1189[g]
Average temperature (January and July, centigrade)	7.3/25.0[e]

a Capital city. b 1990. c 1992. d 1991. e 1980. f 1988. g 1987.

Algeria

Region	Northern Africa
Location (longitude, latitude)	36°75'N 3°05'E
Currency	Algerian dinar
Population (1995 est., in thousands)	28581
Surface area (square kms)	2381741
Population density (pop. per square km)	11
Sex ratio (females per 100 males)	100
Largest city (pop. in thousands)	Algiers (3033)
UN membership date	08-Dec-1962
Major language(s)	Arabic, Berber

Economic indicators	1985	1994
Exchange rate (US$)	4.77	40.90
Consumer price index (1980=100)	155	298[a]
Industrial production index (1980 = 100)	157	169[b]
Balance of payments, current account (million US$)	1015	2367[c]
Tourist arrivals (in thousands)	967[d]	1120[e]
GDP (million US$)	57995	24129[c]
GDP (per capita US$)	2662	941[c]
Long-term rate of change in GDP (% per annum)	5.2	2.3[c]
Gross fixed capital formation (% of GDP)	31.8	...
Economically active female population (%)	6.2[f]	8.2
Economically active male population (%)	74.3[f]	75.2
Annual growth of econ. active female pop. (%)	6.0[b]	...
Annual growth of econ. active male pop. (%)	3.4[b]	...
Agricultural production index (1979-1981=100)	134[d]	179[e]
Food production index (1979-1981=100)	133[d]	178[e]
Commercial energy production (1,000 Mt coal equiv)	100905[f]	149880[c]
Motor vehicles (per 1,000 population)	42	48[b]
Telephone lines (per 100 inhabitants)	2.8[g]	3.4[c]

Social indicators	1990/95
Growth rate of population (% per annum)	2.7
Age group 0-14 years (%)	41.4
Age group 60+ years (women and men, %)	2.9/2.2
Life expectancy at birth (women and men, years)	67/65
Infant mortality rate (per 1,000 births)	61
Total fertility rate (births per woman)	5
Contraceptive use (% of currently married women)	51
Urban population (%)	56
Urban population growth rate (% per annum)	4.2
Rural population growth rate (% per annum)	0.9
Foreign-born (1985,%)	0.9
Refugees	219300
Government education expenditure (% of GDP)	6[h]
Primary-secondary gross enrolment ratio (f and m, per 100)	72/86
Third-level students (per 100,000 population)	979[d]
Newspaper circulation (per 1,000 population)	51
Television receivers (per 1,000 population)	74
Parliamentary seats (women and men, %)	10/90

Environment	1990/91
Threatened species	29
Forested area (%)	2.0
CO_2 emissions (10,000 Mt)	15064
Energy consumption (1,000 Mt coal equiv)	1044
Precipitation (mm)	691[g]
Average temperature (January and July, centigrade)	10.3/24.4[f]

a May 1994. b 1990. c 1991. d 1988. e 1992. f 1980. g 1987. h 1989.

American Samoa

Region	Oceania-Polynesia
Location (longitude, latitude)	14°16'S 170°42'W[a]
Currency	US dollar
Surface area (square kms)	199
Population density (pop. per square km)	191
Largest city (pop. in thousands)	Pago Pago (15)[a]
Major language(s)	English, Samoan

Economic indicators	1985	1994
Consumer price index (1980=100)	104	130[b]
Tourist arrivals (in thousands)	39[c]	31[d]
Motor vehicles (per 1,000 population)	125	...
Telephone lines (per 100 inhabitants)	11.7[b]	12.2[f]

Social indicators	1990/95
Urban population (%)	49
Urban population growth rate (% per annum)	3.8
Rural population growth rate (% per annum)	2.1
Foreign-born (1985,%)	43.9
Newspaper circulation (per 1,000 population)	226[e]
Television receivers (per 1,000 population)	213

Environment	1990/91
Threatened species	13
Forested area (%)	70.0
CO_2 emissions (10,000 Mt)	78
Energy consumption (1,000 Mt coal equiv)	2796

a Capital city. b 1990. c 1988. d 1992. e 1980. f 1991.

Angola

Region Middle Africa
Location (longitude, latitude) 8°82'S 13°24'E
Currency kwanza
Population (1995 est., in thousands) 11072
Surface area (square kms) 1246700
Population density (pop. per square km) 8
Sex ratio (females per 100 males) 103
Largest city (pop. in thousands) Luanda (1642)
UN membership date 01-Dec-1976
Major language(s) Portuguese, Umbundu, Mbundu, Congo

Economic indicators	1985	1994
Exchange rate (US$)	29.92	29.92[a]
Consumer price index (1980=100)	...	733
Balance of payments, current account (million US$)	195	−735[b]
Tourist arrivals (in thousands)	39[c]	40[b]
GDP (million US$)	6865	5954[d]
GDP (per capita US$)	861	625[d]
Long-term rate of change in GDP (% per annum)	3.5	4.5[d]
Economically active female population (%)	56.5[e]	50.2
Economically active male population (%)	88.3[e]	86.0
Annual growth of econ. active female pop. (%)	2.1[a]	...
Annual growth of econ. active male pop. (%)	2.6[a]	...
Labour force in industry (%)	21.8[f]	23.9[g]
Labour force in agriculture (%)	23.7[f]	20.5[g]
Agricultural production index (1979-1981=100)	104[c]	109[b]
Food production index (1979-1981=100)	108[c]	113[b]
Commercial energy production (1,000 Mt coal equiv)	10847[e]	35713[d]
Motor vehicles (per 1,000 population)	21	18[a]
Telephone lines (per 100 inhabitants)	0.7[h]	0.8[d]

Social indicators	1990/95
Growth rate of population (% per annum)	3.7
Age group 0-14 years (%)	47.1
Age group 60+ years (women and men, %)	2.5/2.1
Life expectancy at birth (women and men, years)	48/45
Infant mortality rate (per 1,000 births)	124
Total fertility rate (births per woman)	7
Urban population (%)	32
Urban population growth rate (% per annum)	6.3
Rural population growth rate (% per annum)	2.6
Foreign-born (1985,%)	1.4
Refugees	11000
Third-level students (per 100,000 population)	62[i]
Newspaper circulation (per 1,000 population)	13
Television receivers (per 1,000 population)	6
Parliamentary seats (women and men, %)	10/90

Environment	1990/91
Threatened species	42
Forested area (%)	41.7
CO_2 emissions (10,000 Mt)	1307
Energy consumption (1,000 Mt coal equiv)	93
Precipitation (mm)	367[h]
Average temperature (January and July, centigrade)	25.9/20.2[e]

a 1990. b 1992. c 1988. d 1991. e 1980. f 1983. g 1986. h 1987. i 1989.

Antigua and Barbuda

Region	Caribbean
Location (longitude, latitude)	17°96'N 61°80'W[a]
Currency	East Caribbean dollar
Surface area (square kms)	440
Population density (pop. per square km)	173
Largest city (pop. in thousands)	St. John's (21)[a]
UN membership date	11-Nov-1981
Major language(s)	English

Economic indicators	1985	1994
Exchange rate (US$)	2.70	2.70
Consumer price index (1980=100)	173	213[b]
Balance of payments, current account (million US$)	–23	–9[c]
Tourist arrivals (in thousands)	187[d]	210[c]
GDP (million US$)	200	423[e]
GDP (per capita US$)	3179	6404[e]
Long-term rate of change in GDP (% per annum)	7.8	1.5[e]
Gross fixed capital formation (% of GDP)	28.0	...
Agricultural production index (1979-1981=100)	117[d]	111[c]
Food production index (1979-1981=100)	116[d]	111[c]
Motor vehicles (per 1,000 population)	171	299[e]
Telephone lines (per 100 inhabitants)	14.8[g]	28.8[e]

Social indicators	1990/95
Infant mortality rate (per 1,000 births)	24[h]
Contraceptive use (% of currently married women)	53[d]
Urban population (%)	35
Urban population growth rate (% per annum)	2.2
Rural population growth rate (% per annum)	–0.5
Foreign-born (1985,%)	11.9
Government education expenditure (% of GDP)	4[d]
Newspaper circulation (per 1,000 population)	92
Television receivers (per 1,000 population)	355
Intentional homicides (1986, per 100,000 population)	5
Parliamentary seats (women and men, %)	0/100

Environment	1990/91
Threatened species	6
Forested area (%)	11.4
CO_2 emissions (10,000 Mt)	79
Energy consumption (1,000 Mt coal equiv)	2091

a Capital city. b 1990. c 1992. d 1988. e 1991. f 1980. g 1987. h 1985.

Argentina

Region	South America
Location (longitude, latitude)	34°36'S 58°15'W
Currency	peso
Population (1995 est., in thousands)	34264
Surface area (square kms)	2766889
Population density (pop. per square km)	12
Sex ratio (females per 100 males)	102
Largest city (pop. in thousands)	Buenos Aires (11448)
UN membership date	24-Nov-1945
Major language(s)	Spanish

Economic indicators	1985	1994
Exchange rate (US$)	0.00	1.00
Consumer price index (1980=100)	5	312092[a]
Industrial production index (1980 = 100)	86	99[b]
Balance of payments, current account (million US$)	−952	−7568[c]
Tourist arrivals (in thousands)	2119[d]	3031[e]
GDP (million US$)	88123	189700[b]
GDP (per capita US$)	2905	5799[b]
Long-term rate of change in GDP (% per annum)	−6.6	8.9[b]
Gross fixed capital formation (% of GDP)	14.6[b]	...
Economically active female population (%)	27.3[f]	28.3
Economically active male population (%)	76.3[f]	73.7
Annual growth of econ. active female pop. (%)	1.8[g]	...
Annual growth of econ. active male pop. (%)	0.9[g]	...
Labour force in industry (%)	29.3[h]	30.9[i]
Labour force in agriculture (%)	0.1[h]	0.3[i]
Agricultural production index (1979-1981=100)	112[d]	117[e]
Food production index (1979-1981=100)	112[d]	117[e]
Commercial energy production (1,000 Mt coal equiv)	51040[f]	73187[b]
Motor vehicles (per 1,000 population)	175	179[g]
Telephone lines (per 100 inhabitants)	8.7[i]	9.8[b]

Largest export industries		Major trading partners			1992
	(% of exports)		(% of exports)		(% of imports)
Food, beverages, tobacco	32	Brazil	14	Brazil	23
Agriculture	30	United States	11	United States	22
Chemicals	12	Netherlands	10	Germany	7

Social indicators	1990/95
Growth rate of population (% per annum)	1.2
Age group 0-14 years (%)	28.3
Age group 60+ years (women and men, %)	7.6/5.8
Life expectancy at birth (women and men, years)	75/68
Infant mortality rate (per 1,000 births)	29
Total fertility rate (births per woman)	3
Contraceptive use (% of currently married women)	74[k]
Urban population (%)	87
Urban population growth rate (% per annum)	1.5
Rural population growth rate (% per annum)	−0.9
Foreign-born (1985, %)	6.0
Refugees	11500
Government education expenditure (% of GDP)	2
Primary-secondary gross enrolment ratio (f and m, per 100)	99/93[d]
Third-level students (per 100,000 population)	3079[i]
Newspaper circulation (per 1,000 population)	124
Television receivers (per 1,000 population)	220
Intentional homicides (1986, per 100,000 population)	0
Parliamentary seats (women and men, %)	5/95

Environment	1990/91
Threatened species	86
Forested area (%)	21.4
CO_2 emissions (10,000 Mt)	31618
Energy consumption (1,000 Mt coal equiv)	1977
Precipitation (mm)	1027[i]
Average temperature (January and July, centigrade)	23.7/10.6[f]

a July 1994. b 1991. c 1993. d 1988. e 1992. f 1980. g 1990. h 1982.
i 1984. j 1987. k Source: UNICEF94.

Armenia

Region	Western Asia	
Location (longitude, latitude)	40°10'N 44°31'E[a]	
Currency	rouble	
Surface area (square kms)	29800	
Largest city (pop. in thousands)	Erevan (1210)	
UN membership date	02-Mar-1992	
Major language(s)	Armenian	

Economic indicators	1985	1994
Exchange rate (US$)	...	360.00
GDP (million US$)	9531	8986[b]
GDP (per capita US$)	2962	2632[b]
Long-term rate of change in GDP (% per annum)	6.5	−16.0[b]
Economically active female population (%)	54.6[c]	58.4
Economically active male population (%)	69.5[c]	66.3
Telephone lines (per 100 inhabitants)	14.5[d]	17.7[b]

Social indicators	1990/95
Contraceptive use (% of currently married women)	22
Refugees	300000

Environment	1990/91
Threatened species	28

a Capital city. b 1991. c 1980. d 1987.

Australia

Region	Oceania
Location (longitude, latitude)	33°86'S 151°21'E
Currency	Australian dollar
Population (1995 est., in thousands)	18338
Surface area (square kms)	7713364
Population density (pop. per square km)	2
Sex ratio (females per 100 males)	100
Largest city (pop. in thousands)	Sydney (3671)
UN membership date	01-Nov-1945
Major language(s)	English

Economic indicators	1985	1994
Exchange rate (US$)	1.47	1.35
Consumer price index (1980=100)	149	236
Industrial production index (1980 = 100)	114	147[a]
Unemployment (%)	...	9.3[b]
Balance of payments, current account (million US$)	−8693	−10843[c]
Tourist arrivals (in thousands)	2249[d]	2603[e]
GDP (million US$)	168349	300953[f]
GDP (per capita US$)	10683	17351[f]
Long-term rate of change in GDP (% per annum)	3.9	0.4[f]
Gross fixed capital formation (% of GDP)	24.6	19.6[f]
Economically active female population (%)	45.7[g]	46.6
Economically active male population (%)	77.5[g]	76.6
Annual growth of econ. active female pop. (%)	3.0[h]	...
Annual growth of econ. active male pop. (%)	1.5[h]	...
Labour force in industry (%)	31.0[g]	23.8[e]
Labour force in agriculture (%)	6.5[g]	5.3[e]
Agricultural production index (1979-1981=100)	115[d]	126[e]
Food production index (1979-1981=100)	109[d]	122[e]
Commercial energy production (1,000 Mt coal equiv)	113770[g]	218465[f]
Motor vehicles (per 1,000 population)	532	572[f]
Telephone lines (per 100 inhabitants)	41.9[i]	46.4[f]

Largest export industries		Major trading partners		1992	
	(% of exports)		(% of exports)		(% of imports)
Mining, quarrying	24	Japan	25	United States	23
Other manufacturing	23	United States	11	Japan	18
Metal manufacture	12	Korea, Rep.	6	United Kingdom	6

Social indicators	1990/95
Growth rate of population (% per annum)	1.4
Age group 0-14 years (%)	21.7
Age group 60+ years (women and men, %)	8.5/6.9
Life expectancy at birth (women and men, years)	80/74
Infant mortality rate (per 1,000 births)	7
Total fertility rate (births per woman)	2
Contraceptive use (% of currently married women)	76[i]
Urban population (%)	85
Urban population growth rate (% per annum)	1.4
Rural population growth rate (% per annum)	1.4
Foreign-born (1985, %)	20.3
Refugees	35600
Government education expenditure (% of GDP)	5
Primary-secondary gross enrolment ratio (f and m, per 100)	95/93
Third-level students (per 100,000 population)	2875
Newspaper circulation (per 1,000 population)	246
Television receivers (per 1,000 population)	480
Intentional homicides (1986, per 100,000 population)	2
Parliamentary seats (women and men, %)	8/92

Environment	1990/91
Threatened species	641
Forested area (%)	13.7
CO_2 emissions (10,000 Mt)	71457
Energy consumption (1,000 Mt coal equiv)	7321
Precipitation (mm)	1205[i]
Average temperature (January and July, centigrade)	21.9/12.3[g]

a June 1994. b Labour force sample surveys. c 1993. d 1988. e 1992.
f 1991. g 1980. h 1990. i 1987. j 1986.

Austria

Region	Western Europe
Location (longitude, latitude)	48°21'N 16°36'E
Currency	schilling
Population (1995 est., in thousands)	7861
Surface area (square kms)	83853
Population density (pop. per square km)	93
Sex ratio (females per 100 males)	107
Largest city (pop. in thousands)	Vienna (2122)
UN membership date	14-Dec-1955
Major language(s)	German

Economic indicators	1985	1994
Exchange rate (US$)	17.28	10.90
Consumer price index (1980=100)	127	163
Industrial production index (1980 = 100)	109	115a
Unemployment (%)	...	5.3
Balance of payments, current account (million US$)	−175	−875b
Tourist arrivals (in thousands)	16571c	19098d
GDP (million US$)	65173	163990e
GDP (per capita US$)	8623	21176e
Long-term rate of change in GDP (% per annum)	2.5	3.0e
Gross fixed capital formation (% of GDP)	22.6	24.6f
Economically active female population (%)	42.5g	44.6
Economically active male population (%)	72.7g	74.3
Annual growth of econ. active female pop. (%)	0.9f	...
Annual growth of econ. active male pop. (%)	0.6f	...
Labour force in industry (%)	43.9g	35.2d
Labour force in agriculture (%)	1.5g	0.9d
Agricultural production index (1979-1981=100)	115c	110d
Food production index (1979-1981=100)	115c	110d
Commercial energy production (1,000 Mt coal equiv)	9614g	8337e
Motor vehicles (per 1,000 population)	412	487e
Telephone lines (per 100 inhabitants)	38.2h	43.2e

Largest export industries		Major trading partners		1992
	(% of exports)	(% of exports)		(% of imports)
Metal manufacture	50	Germany	40	Germany 43
Chemicals	13	Italy	9	Italy 9
Textiles	9	Switzerlandi	6	Japan 5

Social indicators	1990/95
Growth rate of population (% per annum)	0.4
Age group 0-14 years (%)	17.5
Age group 60+ years (women and men, %)	12.3/7.9
Life expectancy at birth (women and men, years)	79/73
Infant mortality rate (per 1,000 births)	8
Total fertility rate (births per woman)	2
Urban population (%)	61
Urban population growth rate (% per annum)	1.1
Rural population growth rate (% per annum)	−0.7
Foreign-born (1985,%)	3.6
Refugees	60900
Government education expenditure (% of GDP)	5
Primary-secondary gross enrolment ratio (f and m, per 100)	91/89
Third-level students (per 100,000 population)	2714
Newspaper circulation (per 1,000 population)	351
Television receivers (per 1,000 population)	478
Intentional homicides (1986, per 100,000 population)	2
Parliamentary seats (women and men, %)	21/79

Environment	1990/91
Threatened species	80
Forested area (%)	38.5
CO_2 emissions (10,000 Mt)	16466
Energy consumption (1,000 Mt coal equiv)	4195
Precipitation (mm)	660h
Average temperature (January and July, centigrade)	−1.4/19.9g

a August 1994. b 1993. c 1988. d 1992. e 1991. f 1990. g 1980. h 1987.
i Includes Liechtenstein.

Azerbaijan

Region Western Asia
Location (longitude, latitude) 40°44'N 49°84'E
Currency rouble
Surface area (square kms) 86600
Largest city (pop. in thousands) Baku (1751)
UN membership date 02-Mar-1992
Major language(s) Azerbaijani, Russian

Economic indicators	1985	1994
Exchange rate (US$)	...	1190.00
GDP (million US$)	16905	15310[a]
GDP (per capita US$)	2534	2127[a]
Long-term rate of change in GDP (% per annum)	4.6	−13.8[a]
Economically active female population (%)	53.8[b]	55.5
Economically active male population (%)	75.7[b]	79.6
Telephone lines (per 100 inhabitants)	7.7[c]	9.0[a]

Social indicators	1990/95
Contraceptive use (% of currently married women)	17
Refugees	246000

Environment	1990/91
Threatened species	31

a 1991. b 1980. c 1987.

Bahamas

Region	Caribbean
Location (longitude, latitude)	25°06'N 77°34'W[a]
Currency	Bahamian dollar
Population (1995 est., in thousands)	277
Surface area (square kms)	13878
Population density (pop. per square km)	19
Sex ratio (females per 100 males)	101
Largest city (pop. in thousands)	Nassau (164)[a]
UN membership date	18-Sep-1983
Major language(s)	English

Economic indicators	1985	1994
Exchange rate (US$)	1.00[b]	1.00[b]
Consumer price index (1980=100)	134	202[c]
Balance of payments, current account (million US$)	−48	−127[d]
Tourist arrivals (in thousands)	1475[e]	1399[d]
GDP (million US$)	2049	3013[f]
GDP (per capita US$)	8794	11588[f]
Long-term rate of change in GDP (% per annum)	13.5	−2.2[f]
Gross fixed capital formation (% of GDP)	23.1[f]	...
Economically active female population (%)	37.8[g]	39.2
Economically active male population (%)	78.5[g]	82.0
Labour force in industry (%)	15.5[h]	14.2[i]
Labour force in agriculture (%)	5.0[h]	4.4
Agricultural production index (1979-1981=100)	111[e]	96[d]
Food production index (1979-1981=100)	111[e]	96[d]
Motor vehicles (per 1,000 population)	269	325[j]
Telephone lines (per 100 inhabitants)	22.3[k]	23.8[f]

Social indicators	1990/95
Growth rate of population (% per annum)	1.7
Age group 0-14 years (%)	27.1
Age group 60+ years (women and men, %)	4.3/2.9
Life expectancy at birth (women and men, years)	76/69
Infant mortality rate (per 1,000 births)	24
Total fertility rate (births per woman)	2
Contraceptive use (% of currently married women)	62[l]
Urban population (%)	66
Urban population growth rate (% per annum)	2.3
Rural population growth rate (% per annum)	0.3
Foreign-born (1985, %)	10.9
Refugees	400
Government education expenditure (% of GDP)	3
Primary-secondary gross enrolment ratio (f and m, per 100)	93/90[h]
Newspaper circulation (per 1,000 population)	137
Television receivers (per 1,000 population)	225
Intentional homicides (1986, per 100,000 population)	29

Environment	1990/91
Threatened species	14
Forested area (%)	23.3
CO_2 emissions (10,000 Mt)	531
Energy consumption (1,000 Mt coal equiv)	3538
Precipitation (mm)	1216[k]
Average temperature (January and July, centigrade)	20.3/27.4[g]

a Capital city. b Fixed rate. c June 1994. d 1992. e 1988. f 1991.
g 1980. h 1986. i 1989. j 1990. k 1987. l Source: UNICEF94.

Bahrain

Region	Western Asia
Location (longitude, latitude)	26°13'N 50°35'E[a]
Currency	Bahraini dinar
Population (1995 est., in thousands)	578
Surface area (square kms)	678
Population density (pop. per square km)	762
Sex ratio (females per 100 males)	75
Largest city (pop. in thousands)	Manama (133)[a]
UN membership date	21-Sep-1971
Major language(s)	Arabic

Economic indicators	1985	1994
Exchange rate (US$)	0.38	0.38
Consumer price index (1980=100)	102	102
Balance of payments, current account (million US$)	39	-993[b]
Tourist arrivals (in thousands)	1171[c]	1419[b]
GDP (million US$)	3705	4250[d]
GDP (per capita US$)	8635	8204[d]
Long-term rate of change in GDP (% per annum)	-2.0	8.1[d]
Gross fixed capital formation (% of GDP)	35.0	27.3[d]
Economically active female population (%)	17.6[f]	17.4
Economically active male population (%)	85.7[f]	88.0
Annual growth of econ. active female pop. (%)	13.0[e]	...
Annual growth of econ. active male pop. (%)	6.5[e]	...
Labour force in industry (%)	58.1[g]	56.2[b]
Labour force in agriculture (%)	2.7[g]	1.1[b]
Commercial energy production (1,000 Mt coal equiv)	7674[f]	10123[d]
Motor vehicles (per 1,000 population)	236	260[d]
Telephone lines (per 100 inhabitants)	16.3[g]	19.4[d]

Largest export industries	Major trading partners			1992
(% of exports)	(% of exports)		(% of imports)	
...	Saudi Arabia	4	United States	8
...	Korea, Rep.	3	United Kingdom	7
...	Japan	2	Japan	7

Social indicators	1990/95
Growth rate of population (% per annum)	2.8
Age group 0-14 years (%)	35.3
Age group 60+ years (women and men, %)	1.9/1.9
Life expectancy at birth (women and men, years)	74/69
Infant mortality rate (per 1,000 births)	12
Total fertility rate (births per woman)	4
Contraceptive use (% of currently married women)	53[h]
Urban population (%)	84
Urban population growth rate (% per annum)	3.1
Rural population growth rate (% per annum)	1.3
Foreign-born (1985,%)	40.1
Government education expenditure (% of GDP)	5
Primary-secondary gross enrolment ratio (f and m, per 100)	98/99
Third-level students (per 100,000 population)	1332
Newspaper circulation (per 1,000 population)	57
Television receivers (per 1,000 population)	415
Intentional homicides (1986, per 100,000 population)	1[f]

Environment	1990/91
Threatened species	6
CO_2 emissions (10,000 Mt)	2743
Energy consumption (1,000 Mt coal equiv)	15608
Precipitation (mm)	76[g]
Average temperature (January and July, centigrade)	17.4/33.8[f]

a Capital city. b 1992. c 1988. d 1991. e 1990. f 1980. g 1987. h 1989.

Bangladesh

Region Southern Asia
Location (longitude, latitude) 23°71'N 90°41'E
Currency taka
Population (1995 est., in thousands) 128251
Surface area (square kms) 143998
Population density (pop. per square km) 825
Sex ratio (females per 100 males) 94
Largest city (pop. in thousands) Dacca (6578)
UN membership date 17-Sep-1974
Major language(s) Bengali

Economic indicators	1985	1994
Exchange rate (US$)	31.00	40.25
Consumer price index (1980=100)	175	319[a]
Industrial production index (1980 = 100)	127	217[b]
Balance of payments, current account (million US$)	−458	197[c]
Tourist arrivals (in thousands)	121[d]	110[e]
GDP (million US$)	16654	24683[f]
GDP (per capita US$)	165	212[f]
Long-term rate of change in GDP (% per annum)	4.3	4.0[f]
Gross fixed capital formation (% of GDP)	9.9	10.2[g]
Economically active female population (%)	6.0[h]	7.6
Economically active male population (%)	87.8[h]	87.1
Annual growth of econ. active female pop. (%)	4.4[g]	...
Annual growth of econ. active male pop. (%)	2.5[g]	...
Labour force in industry (%)	11.0[i]	15.4[j]
Labour force in agriculture (%)	58.8[i]	65.0[j]
Agricultural production index (1979-1981=100)	113[d]	131[e]
Food production index (1979-1981=100)	114[d]	132[e]
Commercial energy production (1,000 Mt coal equiv)	1617[h]	6562[f]
Motor vehicles (per 1,000 population)	1	1[f]
Telephone lines (per 100 inhabitants)	0.2[k]	0.2[f]

Social indicators	1990/95
Growth rate of population (% per annum)	2.4
Age group 0-14 years (%)	40.3
Age group 60+ years (women and men, %)	2.2/2.5
Life expectancy at birth (women and men, years)	53/53
Infant mortality rate (per 1,000 births)	108
Total fertility rate (births per woman)	5
Contraceptive use (% of currently married women)	40
Urban population (%)	20
Urban population growth rate (% per annum)	5.9
Rural population growth rate (% per annum)	1.7
Foreign-born (1985,%)	0.8
Refugees	245000
Government education expenditure (% of GDP)	2
Primary-secondary gross enrolment ratio (f and m, per 100)	37/47
Third-level students (per 100,000 population)	310
Newspaper circulation (per 1,000 population)	6
Television receivers (per 1,000 population)	5
Intentional homicides (1986, per 100,000 population)	3
Parliamentary seats (women and men, %)	10/90

Environment	1990/91
Threatened species	66
Forested area (%)	12.9
CO_2 emissions (10,000 Mt)	4215
Energy consumption (1,000 Mt coal equiv)	77

a June 1994. b April 1994. c 1993. d 1988. e 1992. f 1991. g 1990.
h 1980. i 1983. j 1989. k 1987.

Barbados

Region Caribbean
Location (longitude, latitude) 13°06'N 59°37'W[a]
Currency Barbados dollar
Population (1995 est., in thousands) 261
Surface area (square kms) 430
Population density (pop. per square km) 593
Sex ratio (females per 100 males) 109
Largest city (pop. in thousands) Bridgetown (115)[a]
UN membership date 09-Dec-1966
Major language(s) English

Economic indicators	1985	1994
Exchange rate (US$)	2.01	2.01
Consumer price index (1980=100)	145	200
Industrial production index (1980 = 100)	97	99[b]
Unemployment (%)	...	22.9[c]
Balance of payments, current account (million US$)	40	137[d]
Tourist arrivals (in thousands)	451[e]	385[f]
GDP (million US$)	1198	1687[g]
GDP (per capita US$)	4737	6539[g]
Long-term rate of change in GDP (% per annum)	0.7	−2.8[g]
Gross fixed capital formation (% of GDP)	15.1	16.2[g]
Economically active female population (%)	59.0[h]	61.5
Economically active male population (%)	76.4[h]	78.4
Annual growth of econ. active female pop. (%)	2.9[i]	...
Annual growth of econ. active male pop. (%)	1.4[i]	...
Labour force in industry (%)	21.6[j]	19.7[f]
Labour force in agriculture (%)	9.4[j]	6.2[f]
Agricultural production index (1979-1981=100)	78[e]	77[f]
Food production index (1979-1981=100)	78[e]	77[f]
Commercial energy production (1,000 Mt coal equiv)	78[h]	119[g]
Motor vehicles (per 1,000 population)	149	...
Telephone lines (per 100 inhabitants)	25.6[k]	30.2[g]

Social indicators	1990/95
Growth rate of population (% per annum)	0.3
Age group 0-14 years (%)	23.4
Age group 60+ years (women and men, %)	9.2/6.1
Life expectancy at birth (women and men, years)	78/73
Infant mortality rate (per 1,000 births)	10
Total fertility rate (births per woman)	2
Contraceptive use (% of currently married women)	55[e]
Urban population (%)	48
Urban population growth rate (% per annum)	1.6
Rural population growth rate (% per annum)	−0.8
Foreign-born (1985,%)	7.9
Government education expenditure (% of GDP)	6
Primary-secondary gross enrolment ratio (f and m, per 100)	97/102[l]
Third-level students (per 100,000 population)	1665[l]
Newspaper circulation (per 1,000 population)	117
Television receivers (per 1,000 population)	265
Intentional homicides (1986, per 100,000 population)	6[h]
Parliamentary seats (women and men, %)	4/96

Environment	1990/91
Threatened species	3
CO_2 emissions (10,000 Mt)	276
Energy consumption (1,000 Mt coal equiv)	1601
Precipitation (mm)	1273[k]
Average temperature (January and July, centigrade)	25.2/26.8[h]

a Capital city. b April 1994. c Labour force sample surveys. d 1993.
e 1988. f 1992. g 1991. h 1980. i 1990. j 1981. k 1987. l 1989.

Belarus

Region Eastern Europe
Location (longitude, latitude) 53°90'N 27°55'E
Currency roubles
Surface area (square kms) 207600
Largest city (pop. in thousands) Minsk (1648)
UN membership date 24-Oct-1945
Major language(s) Byelorussian, Russian

Economic indicators	1985	1994
Exchange rate (US$)	...	3400.00
Consumer price index (1980=100)	105	8065[a]
GDP (million US$)	32575	45700[b]
GDP (per capita US$)	3258	4446[b]
Long-term rate of change in GDP (% per annum)	4.0	−3.0[b]
Economically active female population (%)	61.4[c]	59.3
Economically active male population (%)	79.1[c]	79.2
Labour force in industry (%)	39.7[d]	39.9[e]
Labour force in agriculture (%)	22.1[d]	20.7[e]
Telephone lines (per 100 inhabitants)	12.4[d]	16.3[e]

Social indicators	1990/95
Contraceptive use (% of currently married women)	23
Newspaper circulation (per 1,000 population)	286
Television receivers (per 1,000 population)	250[f]
Intentional homicides (1986, per 100,000 population)	7[g]

Environment	1990/91
Threatened species	31

a March 1994. b 1991. c 1980. d 1987. e 1992. f 1985. g Source: DYB92.

Belgium

Region	Western Europe
Location (longitude, latitude)	50°85'N 4°37'E
Currency	Belgian franc
Population (1995 est., in thousands)	10031
Surface area (square kms)	30519
Population density (pop. per square km)	323
Sex ratio (females per 100 males)	104
Largest city (pop. in thousands)	Brussels (1317)
UN membership date	27-Dec-1945
Major language(s)	Dutch, French, German

Economic indicators	1985	1994
Exchange rate (US$)	50.36	31.83
Consumer price index (1980=100)	141	174
Industrial production index (1980 = 100)	104	118[a]
Unemployment (%)	...	14.6
Tourist arrivals (in thousands)	2700[b]	3220[c]
GDP (million US$)	79841	196874[d]
GDP (per capita US$)	8099	19717[d]
Long-term rate of change in GDP (% per annum)	0.8	1.9[d]
Gross fixed capital formation (% of GDP)	15.6	19.8[d]
Economically active female population (%)	32.9[e]	32.8
Economically active male population (%)	68.7[f]	69.3
Annual growth of econ. active female pop. (%)	1.2[f]	...
Annual growth of econ. active male pop. (%)	0.5[f]	...
Labour force in industry (%)	34.3[e]	27.9[d]
Labour force in agriculture (%)	3.1[e]	2.6[d]
Agricultural production index (1979-1981=100)[g]	113[b]	136[c]
Food production index (1979-1981=100)[g]	113[b]	136[c]
Commercial energy production (1,000 Mt coal equiv)	10537[e]	16603[d]
Motor vehicles (per 1,000 population)	367	436[d]
Telephone lines (per 100 inhabitants)	34.4[h]	41.0[d]

Largest export industries	Major trading partners			1992
(% of exports)[g]	(% of exports)[g]		(% of imports)[g]	
Metal manufacture 32	Germany	23	Germany	24
Chemicals 21	France[i]	19	Netherlands	18
Food, beverages, tobacco 9	Netherlands	14	France[i]	17

Social indicators	1990/95
Growth rate of population (% per annum)	0.1
Age group 0-14 years (%)	18.0
Age group 60+ years (women and men, %)	12.2/8.9
Life expectancy at birth (women and men, years)	79/73
Infant mortality rate (per 1,000 births)	8
Total fertility rate (births per woman)	2
Contraceptive use (% of currently married women)	79
Urban population (%)	97
Urban population growth rate (% per annum)	0.2
Rural population growth rate (% per annum)	-1.8
Foreign-born (1985, %)	9.0
Refugees	24300
Government education expenditure (% of GDP)	5
Primary-secondary gross enrolment ratio (f and m, per 100)	103/103
Third-level students (per 100,000 population)	2754[j]
Newspaper circulation (per 1,000 population)	301
Television receivers (per 1,000 population)	451
Intentional homicides (1986, per 100,000 population)	2[k]
Parliamentary seats (women and men, %)	9/91

Environment	1990/91
Threatened species	39
Forested area (%)[g]	21.1
CO_2 emissions (10,000 Mt)	27860
Energy consumption (1,000 Mt coal equiv)	6900
Precipitation (mm)	766[l]
Average temperature (January and July, centigrade)	2.2/17.5[e]

a January 1994. b 1988. c 1992. d 1991. e 1980. f 1990. g Includes Luxemburg. h 1987. i Includes Monaco. j 1989. k 1984. l Uccle.

Belize

Region Central America
Location (longitude, latitude) 17°30'N 88°12'W[a]
Currency Belize dollar
Population (1995 est., in thousands) 183
Surface area (square kms) 22965
Population density (pop. per square km) 8
Sex ratio (females per 100 males) 97
Largest city (pop. in thousands) Belize City (44)[a]
UN membership date 25-Sep-1981
Major language(s) English, Spanish

Economic indicators	1985	1994
Exchange rate (US$)	2.00	2.00
Consumer price index (1980=100)	134	167[b]
Balance of payments, current account (million US$)	9	−49[c]
Tourist arrivals (in thousands)	142[d]	247[e]
GDP (million US$)	209	420[f]
GDP (per capita US$)	1260	2165[f]
Long-term rate of change in GDP (% per annum)	1.0	5.4[f]
Gross fixed capital formation (% of GDP)	17.4	29.6[f]
Economically active female population (%)	27.8[g]	28.8
Economically active male population (%)	82.3[g]	80.7
Agricultural production index (1979-1981=100)	107[d]	122[e]
Food production index (1979-1981=100)	107[d]	122[e]
Motor vehicles (per 1,000 population)	39	27[f]
Telephone lines (per 100 inhabitants)	5.7[h]	10.4[f]

Largest export industries	Major trading partners		1992
(% of exports)	(% of exports)	(% of imports)	
Food, beverages, tobacco 59	United States 47	United States 57	
Textiles 14	United Kingdom 24	Mexico 9	
Agriculture 14	Mexico 13	United Kingdom 9	

Social indicators	1990/95
Growth rate of population (% per annum)	2.5
Age group 0-14 years (%)	44.6
Age group 60+ years (women and men, %)	3.9/3.7[i][j]
Life expectancy at birth (women and men, years)	72/70[k]
Infant mortality rate (per 1,000 births)	21
Contraceptive use (% of currently married women)	47
Urban population (%)	53
Urban population growth rate (% per annum)	2.8
Rural population growth rate (% per annum)	1.2
Foreign-born (1985,%)	11.0
Refugees	20400
Government education expenditure (% of GDP)	5
Newspaper circulation (per 1,000 population)	18[g]
Television receivers (per 1,000 population)	165
Intentional homicides (1986, per 100,000 population)	16
Parliamentary seats (women and men, %)	0/100

Environment	1990/91
Threatened species	15
Forested area (%)	44.1
CO_2 emissions (10,000 Mt)	72
Energy consumption (1,000 Mt coal equiv)	655
Precipitation (mm)	1648[h]
Average temperature (January and July, centigrade)	23.5/27.6[g]

a Capital city. b November 1993. c 1993. d 1988. e 1992. f 1991.
g 1980. h 1987. i Source DYB90. j 1989. k 1975-80 data.

Benin

Region	Southern Africa
Location (longitude, latitude)	6°21'N 2°26'E
Currency	CFA franc
Population (1995 est., in thousands)	5399
Surface area (square kms)	112622
Population density (pop. per square km)	43
Sex ratio (females per 100 males)	102
Largest city (pop. in thousands)	Cotonou (787)
UN membership date	20-Sep-1960
Major language(s)	French, Fon

Economic indicators	1985	1994
Exchange rate (US$)	378.05	528.15
Consumer price index (1980=100)	...	104
Balance of payments, current account (million US$)	5	−29[a]
Tourist arrivals (in thousands)	75[b]	130[a]
GDP (million US$)	1046	1849[c]
GDP (per capita US$)	263	388[c]
Long-term rate of change in GDP (% per annum)	7.5	2.1[c]
Gross fixed capital formation (% of GDP)	13.4	...
Economically active female population (%)	83.5[d]	74.8
Economically active male population (%)	90.2[d]	88.3
Annual growth of econ. active female pop. (%)	2.1[e]	...
Annual growth of econ. active male pop. (%)	2.2[e]	...
Labour force in industry (%)	22.2[d]	18.9[f]
Labour force in agriculture (%)	7.6[d]	7.5[f]
Agricultural production index (1979-1981=100)	155[b]	185[a]
Food production index (1979-1981=100)	149[b]	177[a]
Commercial energy production (1,000 Mt coal equiv)	14[d]	421[c]
Motor vehicles (per 1,000 population)	9	7[e]
Telephone lines (per 100 inhabitants)	0.3[g]	0.3[c]

Social indicators	1990/95
Growth rate of population (% per annum)	3.1
Age group 0-14 years (%)	47.3
Age group 60+ years (women and men, %)	2.4/2.1
Life expectancy at birth (women and men, years)	48/45
Infant mortality rate (per 1,000 births)	87
Total fertility rate (births per woman)	7
Urban population (%)	42
Urban population growth rate (% per annum)	4.9
Rural population growth rate (% per annum)	1.9
Foreign-born (1985,%)	1.1
Refugees	300
Government education expenditure (% of GDP)	4[f]
Primary-secondary gross enrolment ratio (f and m, per 100)	28/58[f]
Third-level students (per 100,000 population)	235
Newspaper circulation (per 1,000 population)	3
Television receivers (per 1,000 population)	5
Parliamentary seats (women and men, %)	6/94

Environment	1990/91
Threatened species	12
Forested area (%)	30.8
CO_2 emissions (10,000 Mt)	153
Energy consumption (1,000 Mt coal equiv)	48
Precipitation (mm)	1339[g]
Average temperature (January and July, centigrade)	27.1/25.2[d]

a 1992. b 1988. c 1991. d 1980. e 1990. f 1985. g 1987.

Bhutan

Region	Southern Asia
Location (longitude, latitude)	27°57'N 89°65'E[a]
Currency	ngultrum
Population (1995 est., in thousands)	1729
Surface area (square kms)	47000
Population density (pop. per square km)	33
Sex ratio (females per 100 males)	98
Largest city (pop. in thousands)	Thimphu (17)[a]
UN membership date	21-Sep-1971
Major language(s)	Dzongkha, Assamese, Nepali

Economic indicators	1985	1994
Exchange rate (US$)	12.16	31.30
Balance of payments, current account (million US$)	−16	...
Tourist arrivals (in thousands)	2[b]	3[c]
GDP (million US$)	193	246[d]
GDP (per capita US$)	140	156[d]
Long-term rate of change in GDP (% per annum)	3.7	5.0[d]
Gross fixed capital formation (% of GDP)	41.9	34.1[d]
Economically active female population (%)	45.2[f]	41.7
Economically active male population (%)	89.9[f]	89.1
Annual growth of econ. active female pop. (%)	1.5[e]	...
Annual growth of econ. active male pop. (%)	2.1[e]	...
Agricultural production index (1979-1981=100)	97[b]	111[c]
Food production index (1979-1981=100)	97[b]	110[c]
Commercial energy production (1,000 Mt coal equiv)	1[f]	195[d]
Telephone lines (per 100 inhabitants)	0.1[g]	0.2[d]

Social indicators	1990/95
Growth rate of population (% per annum)	2.3
Age group 0-14 years (%)	40.8
Age group 60+ years (women and men, %)	2.9/2.5
Life expectancy at birth (women and men, years)	49/48
Infant mortality rate (per 1,000 births)	129
Total fertility rate (births per woman)	6
Contraceptive use (% of currently married women)	2[h]
Urban population (%)	6
Urban population growth rate (% per annum)	6.0
Rural population growth rate (% per annum)	2.1
Foreign-born (1985,%)	0.6
Primary-secondary gross enrolment ratio (f and m, per 100)	13/22[b]
Third-level students (per 100,000 population)	26[f]
Parliamentary seats (women and men, %)	0/100

Environment	1990/91
Threatened species	41
Forested area (%)	55.5
CO_2 emissions (10,000 Mt)	35
Energy consumption (1,000 Mt coal equiv)	50

a Capital city. b 1988. c 1992. d 1991. e 1990. f 1980. g 1987. h Source: UNICEF94.

Bolivia

Region South America
Location (longitude, latitude) 16°49'S 68°15'W
Currency boliviano
Population (1995 est., in thousands) 8074
Surface area (square kms) 1098581
Population density (pop. per square km) 7
Sex ratio (females per 100 males) 102
Largest city (pop. in thousands) La Paz (1010)
UN membership date 14-Nov-1945
Major language(s) Spanish, Quechua, Aymara

Economic indicators	1985	1994
Exchange rate (US$)	1.69	4.66
Consumer price index (1980=100)[a]	18158	128[bc]
Industrial production index (1980 = 100)	66	93
Balance of payments, current account (million US$)	−282	−533[d]
Tourist arrivals (in thousands)	167[e]	245[d]
GDP (million US$)	6487	6058[f]
GDP (per capita US$)	1023	825[f]
Long-term rate of change in GDP (% per annum)	−1.0	4.1[f]
Gross fixed capital formation (% of GDP)	11.7	12.0[f]
Economically active female population (%)	22.5[g]	25.9
Economically active male population (%)	85.0[g]	82.4
Annual growth of econ. active female pop. (%)	3.5[h]	...
Annual growth of econ. active male pop. (%)	2.3[h]	...
Labour force in industry (%)	20.6[g]	12.8[h]
Labour force in agriculture (%)	46.5[g]	47.4[h]
Agricultural production index (1979-1981=100)	118[e]	136[d]
Food production index (1979-1981=100)	120[e]	137[d]
Commercial energy production (1,000 Mt coal equiv)	4790[g]	5605[f]
Motor vehicles (per 1,000 population)	13	29[f]
Telephone lines (per 100 inhabitants)	2.4[i]	2.5[f]

Largest export industries	Major trading partners		1992
(% of exports)	(% of exports)	(% of imports)	
Mining, quarrying 53	Argentina 20	United States 23	
Basic metal industry 13	United States 20	Brazil 15	
Food, beverages, tobacco 9	United Kingdom 17	Japan 12	

Social indicators	1990/95
Growth rate of population (% per annum)	2.4
Age group 0-14 years (%)	39.7
Age group 60+ years (women and men, %)	3.3/2.7
Life expectancy at birth (women and men, years)	64/59
Infant mortality rate (per 1,000 births)	85
Total fertility rate (births per woman)	5
Contraceptive use (% of currently married women)	30[i]
Urban population (%)	54
Urban population growth rate (% per annum)	3.7
Rural population growth rate (% per annum)	0.9
Foreign-born (1985,%)	1.1
Refugees	500
Government education expenditure (% of GDP)	2[i]
Primary-secondary gross enrolment ratio (f and m, per 100)	64/72
Third-level students (per 100,000 population)	1980[j]
Newspaper circulation (per 1,000 population)	56
Television receivers (per 1,000 population)	103

Environment	1990/91
Threatened species	58
Forested area (%)	50.6
CO_2 emissions (10,000 Mt)	1598
Energy consumption (1,000 Mt coal equiv)	374
Precipitation (mm)	105[k]
Average temperature (January and July, centigrade)	25.4/19.7[g]

a Multiply each figure by 100. b March 1994. c Index base: 1991 = 100.
d 1992. e 1988. f 1991. g 1980. h 1990. i 1987. j 1989. k Santa Cruz.

Bosnia-Herzegovina

Region Southern Europe
Location (longitude, latitude) 43°52'N 18°26'E
Currency new dinar
Largest city (pop. in thousands) Sarajevo (416)[a]
UN membership date 22-May-1992
Major language(s) Serbo-Croatian

Economic indicators	1985	1994
GDP (million US$)	5672	17631[b]
GDP (per capita US$)	1314	3639[b]
Long-term rate of change in GDP (% per annum)	1.5	−29.9[b]
Economically active female population (%)	42.8[c]	43.3
Economically active male population (%)	74.9[c]	75.2

Social indicators	1990/95
Refugees	810000

a Capital city. b 1991. c 1980.

Botswana

Region	Southern Africa
Location (longitude, latitude)	24°45'S 25°55'E[a]
Currency	pula
Population (1995 est., in thousands)	1433
Surface area (square kms)	581730
Population density (pop. per square km)	2
Sex ratio (females per 100 males)	108
Largest city (pop. in thousands)	Gaborone (97)[a]
UN membership date	17-Oct-1966
Major language(s)	Tswana, English

Economic indicators	1985	1994
Exchange rate (US$)	2.10	2.82
Consumer price index (1980=100)	144	393
Balance of payments, current account (million US$)	81	138[b]
Tourist arrivals (in thousands)	268[c]	437[d]
GDP (million US$)	968	3467[e]
GDP (per capita US$)	907	2720[e]
Long-term rate of change in GDP (% per annum)	7.2	8.3[e]
Gross fixed capital formation (% of GDP)	18.9	...
Economically active female population (%)	46.0[f]	41.3
Economically active male population (%)	85.3[f]	85.1
Annual growth of econ. active female pop. (%)	2.2[b]	...
Annual growth of econ. active male pop. (%)	4.4[b]	...
Labour force in industry (%)	33.0[f]	30.6[d]
Labour force in agriculture (%)	5.0[f]	2.6[d]
Agricultural production index (1979-1981=100)	107[c]	93[d]
Food production index (1979-1981=100)	107[c]	93[d]
Motor vehicles (per 1,000 population)	40	58[e]
Telephone lines (per 100 inhabitants)	1.3[g]	2.6[e]

Social indicators	1990/95
Growth rate of population (% per annum)	2.9
Age group 0-14 years (%)	44.9
Age group 60+ years (women and men, %)	2.8/2.2
Life expectancy at birth (women and men, years)	64/58
Infant mortality rate (per 1,000 births)	60
Total fertility rate (births per woman)	5
Contraceptive use (% of currently married women)	33[c]
Urban population (%)	31
Urban population growth rate (% per annum)	7.1
Rural population growth rate (% per annum)	1.3
Foreign-born (1985,%)	2.2
Refugees	500
Government education expenditure (% of GDP)	7
Primary-secondary gross enrolment ratio (f and m, per 100)	90/85
Third-level students (per 100,000 population)	255[h]
Newspaper circulation (per 1,000 population)	15
Television receivers (per 1,000 population)	16
Parliamentary seats (women and men, %)	5/95

Environment	1990/91
Threatened species	14
Forested area (%)	18.8
CO_2 emissions (10,000 Mt)	588

a Capital city. b 1990. c 1988. d 1992. e 1991. f 1980. g 1987. h 1989.

Brazil

Region	South America
Location (longitude, latitude)	23°55'S 46°63'W
Currency	cruzeiro
Population (1995 est., in thousands)	161382
Surface area (square kms)	8511965
Population density (pop. per square km)	18
Sex ratio (females per 100 males)	101
Largest city (pop. in thousands)	Sao Paulo (18119)
UN membership date	24-Oct-1945
Major language(s)	Portuguese

Economic indicators	1985	1994
Exchange rate (US$)	0.10	0.85[a]
Consumer price index (1980=100)[b]	7	4136690[c]
Industrial production index (1980 = 100)	99	110[d]
Unemployment (%)	...	5.9[e]
Balance of payments, current account (million US$)	−273	6275[f]
Tourist arrivals (in thousands)	1743[g]	1475[f]
GDP (million US$)	223065	401091[h]
GDP (per capita US$)	1645	2646[h]
Long-term rate of change in GDP (% per annum)	8.3	0.9[h]
Gross fixed capital formation (% of GDP)	16.9	...
Economically active female population (%)	30.0[i]	30.8
Economically active male population (%)	82.2[i]	80.2
Annual growth of econ. active female pop. (%)	4.3[j]	...
Annual growth of econ. active male pop. (%)	2.6[j]	...
Labour force in industry (%)	24.7[k]	22.7[i]
Labour force in agriculture (%)	29.3[k]	22.8[i]
Agricultural production index (1979-1981=100)	131[g]	140[f]
Food production index (1979-1981=100)	136[g]	146[f]
Commercial energy production (1,000 Mt coal equiv)	34056[i]	83017[h]
Motor vehicles (per 1,000 population)	88	88[i]
Telephone lines (per 100 inhabitants)	5.6[l]	6.6[h]

Largest export industries		Major trading partners			1992
	(% of exports)		(% of exports)		(% of imports)
Metal manufacture	24	United States	20	United States	23
Food, beverages, tobacco	17	Argentina	9	Germany	9
Agriculture	9	Netherlands	7	Argentina	8

Social indicators	1990/95
Growth rate of population (% per annum)	1.6
Age group 0-14 years (%)	32.2
Age group 60+ years (women and men, %)	4.0/3.6
Life expectancy at birth (women and men, years)	69/64
Infant mortality rate (per 1,000 births)	57
Total fertility rate (births per woman)	3
Contraceptive use (% of currently married women)	66[m]
Urban population (%)	79
Urban population growth rate (% per annum)	2.3
Rural population growth rate (% per annum)	−1.5
Foreign-born (1985,%)	0.8
Refugees	5400
Primary-secondary gross enrolment ratio (f and m, per 100)	80/82[i]
Third-level students (per 100,000 population)	1064
Newspaper circulation (per 1,000 population)	54
Television receivers (per 1,000 population)	207
Parliamentary seats (women and men, %)	6/94

Environment	1990/91
Threatened species	204
Forested area (%)	57.9
CO_2 emissions (10,000 Mt)	58843
Energy consumption (1,000 Mt coal equiv)	799
Precipitation (mm)	1386[l]
Average temperature (January and July, centigrade)	22.1/15.0[i]

a Cruzeiro real. b Multiply each figure by 1000. c August 1994. d June 1994. e Labour force sample surveys. f 1992. g 1988. h 1991. i 1980. j 1990. k 1981. l 1987. m 1986.

Brunei Darussalam

Region	South-eastern Asia
Location (longitude, latitude)	4°56'N 114°55'E[a]
Currency	Brunei dollar
Population (1995 est., in thousands)	288
Surface area (square kms)	5765
Population density (pop. per square km)	47
Sex ratio (females per 100 males)	95
Largest city (pop. in thousands)	Bandar Seri Begawan (64)[a]
UN membership date	21-Sep-1984
Major language(s)	Malay, English

Economic indicators	1985	1994
Consumer price index (1980=100)	124	...
Tourist arrivals (in thousands)	457[b]	500[c]
GDP (million US$)	3122	3816[d]
GDP (per capita US$)	13815	14456[d]
Long-term rate of change in GDP (% per annum)	−1.5	3.6[d]
Economically active female population (%)	50.8[e]	48.2
Economically active male population (%)	86.5[e]	86.2
Labour force in industry (%)	61.3[e]	55.7[f]
Labour force in agriculture (%)	1.8[e]	2.3[f]
Agricultural production index (1979-1981=100)	168[b]	141[c]
Food production index (1979-1981=100)	168[b]	142[c]
Commercial energy production (1,000 Mt coal equiv)	29779[e]	24418[d]
Motor vehicles (per 1,000 population)	391	490[d]
Telephone lines (per 100 inhabitants)	10.3[g]	14.8[d]

Social indicators	1990/95
Growth rate of population (% per annum)	2.3
Age group 0-14 years (%)	31.9
Age group 60+ years (women and men, %)	3.8/3.8
Life expectancy at birth (women and men, years)	76/73
Infant mortality rate (per 1,000 births)	8
Total fertility rate (births per woman)	3
Urban population (%)	58
Urban population growth rate (% per annum)	2.2
Rural population growth rate (% per annum)	2.2
Foreign-born (1985, %)	27.8
Primary-secondary gross enrolment ratio (f and m, per 100)	89/88
Newspaper circulation (per 1,000 population)	39
Television receivers (per 1,000 population)	235

Environment	1990/91
Threatened species	37
Forested area (%)	39.0
CO_2 emissions (10,000 Mt)	1530
Energy consumption (1,000 Mt coal equiv)	17178

a Capital city. b 1988. c 1992. d 1991. e 1980. f 1986. g 1987.

Bulgaria

Region	Eastern Europe
Location (longitude, latitude)	42°69'N 23°31'E
Currency	leva
Population (1995 est., in thousands)	8887
Surface area (square kms)	110912
Population density (pop. per square km)	81
Sex ratio (females per 100 males)	104
Largest city (pop. in thousands)	Sofia (1313)
UN membership date	14-Dec-1955
Major language(s)	Bulgarian

Economic indicators	1985	1994
Exchange rate (US$)	1.00	53.90
Consumer price index (1980=100)	105	787
Industrial production index (1980 = 100)	124	91[a]
Balance of payments, current account (million US$)	–136	452[b]
Tourist arrivals (in thousands)	3967[c]	3750[b]
GDP (million US$)	32273	7368[b]
GDP (per capita US$)	3602	821[b]
Long-term rate of change in GDP (% per annum)	1.8	–6.9[b]
Gross fixed capital formation (% of GDP)	26.4	18.9[d]
Economically active female population (%)	61.4[e]	59.8
Economically active male population (%)	70.6[e]	68.2
Annual growth of econ. active female pop. (%)	0.3[f]	...
Annual growth of econ. active male pop. (%)	–0.1[f]	...
Labour force in industry (%)	42.7[e]	39.2[b]
Labour force in agriculture (%)	24.4[e]	18.0[b]
Agricultural production index (1979-1981=100)	98[c]	80[b]
Food production index (1979-1981=100)	100[c]	86[b]
Commercial energy production (1,000 Mt coal equiv)	10992[e]	12299[d]
Motor vehicles (per 1,000 population)	162[f]	...
Telephone lines (per 100 inhabitants)	19.4[g]	24.6[d]

Largest export industries	Major trading partners		1992
(% of exports)	(% of exports)	(% of imports)	
...	...	Russian Fed.	23
...	...	Germany	13
...	...	Italy	5

Social indicators	1990/95
Growth rate of population (% per annum)	–0.2
Age group 0-14 years (%)	19.2
Age group 60+ years (women and men, %)	11.3/9.1
Life expectancy at birth (women and men, years)	75/69
Infant mortality rate (per 1,000 births)	14
Total fertility rate (births per woman)	2
Urban population (%)	71
Urban population growth rate (% per annum)	0.6
Rural population growth rate (% per annum)	–2.2
Foreign-born (1985,%)	0.3
Refugees	200
Government education expenditure (% of GDP)	6
Primary-secondary gross enrolment ratio (f and m, per 100)	88/89
Third-level students (per 100,000 population)	2092
Newspaper circulation (per 1,000 population)	452
Television receivers (per 1,000 population)	252
Intentional homicides (1986, per 100,000 population)	3
Parliamentary seats (women and men, %)	13/87

Environment	1990/91
Threatened species	47
Forested area (%)	34.9
CO_2 emissions (10,000 Mt)	15468
Energy consumption (1,000 Mt coal equiv)	3270
Precipitation (mm)	622[g]
Average temperature (January and July, centigrade)	–1.7/21.3[e]

a August 1994. b 1992. c 1988. d 1991. e 1980. f 1990. g 1987.

Burkina Faso

Region	Southern Africa
Location (longitude, latitude)	12°22'N 1°31'W[a]
Currency	CFA franc
Population (1995 est., in thousands)	10352
Surface area (square kms)	274000
Population density (pop. per square km)	34
Sex ratio (females per 100 males)	102
Largest city (pop. in thousands)	Ouagadougou (681)[a]
UN membership date	20-Sep-1960
Major language(s)	French, Dyula, Ful

Economic indicators	1985	1994
Exchange rate (US$)	378.05	528.15
Consumer price index (1980=100)	112	137[b]
Balance of payments, current account (million US$)	−60	−99[c]
Tourist arrivals (in thousands)	83[d]	74[c]
GDP (million US$)	1045	2221[e]
GDP (per capita US$)	133	240[e]
Long-term rate of change in GDP (% per annum)	10.3	2.0[e]
Gross fixed capital formation (% of GDP)	24.2	...
Economically active female population (%)	83.3[f]	74.9
Economically active male population (%)	93.8[f]	92.8
Annual growth of econ. active female pop. (%)	1.7[g]	...
Annual growth of econ. active male pop. (%)	2.1[g]	...
Agricultural production index (1979-1981=100)	158[d]	188[c]
Food production index (1979-1981=100)	156[d]	186[c]
Motor vehicles (per 1,000 population)	6	7[e]
Telephone lines (per 100 inhabitants)	0.1[h]	0.2[e]

Social indicators	1990/95
Growth rate of population (% per annum)	2.8
Age group 0-14 years (%)	44.9
Age group 60+ years (women and men, %)	2.7/2.3
Life expectancy at birth (women and men, years)	50/47
Infant mortality rate (per 1,000 births)	118
Total fertility rate (births per woman)	7
Contraceptive use (% of currently married women)	8
Urban population (%)	20
Urban population growth rate (% per annum)	7.8
Rural population growth rate (% per annum)	1.8
Foreign-born (1985,%)	3.3
Refugees	5700
Government education expenditure (% of GDP)	2[i]
Primary-secondary gross enrolment ratio (f and m, per 100)	16/27[i]
Third-level students (per 100,000 population)	65[i]
Newspaper circulation (per 1,000 population)	0
Television receivers (per 1,000 population)	5
Parliamentary seats (women and men, %)	6/94

Environment	1990/91
Threatened species	11
Forested area (%)	24.1
CO_2 emissions (10,000 Mt)	152
Energy consumption (1,000 Mt coal equiv)	29
Precipitation (mm)	879[h]
Average temperature (January and July, centigrade)	25.4/26.9[f]

a Capital city. b April 1994. c 1992. d 1988. e 1991. f 1980. g 1990. h 1987. i 1989.

Burundi

Region	Eastern Africa
Location (longitude, latitude)	3°19'S 29°19'E[a]
Currency	Burundi franc
Population (1995 est., in thousands)	6343
Surface area (square kms)	27834
Population density (pop. per square km)	202
Sex ratio (females per 100 males)	104
Largest city (pop. in thousands)	Bujumbura (240)[a]
UN membership date	18-Sep-1962
Major language(s)	Kirundi, French

Economic indicators	1985	1994
Exchange rate (US$)	111.96	248.52
Consumer price index (1980=100)	152	104[b]
Balance of payments, current account (million US$)	−42	−54[c]
Tourist arrivals (in thousands)	99[d]	86[c]
GDP (million US$)	1171	1170[e]
GDP (per capita US$)	247	207[e]
Long-term rate of change in GDP (% per annum)	11.7	1.8[e]
Gross fixed capital formation (% of GDP)	14.2	
Economically active female population (%)	83.2[f]	76.4
Economically active male population (%)	93.1[f]	92.8
Annual growth of econ. active female pop. (%)	1.4[g]	...
Annual growth of econ. active male pop. (%)	1.9[g]	...
Labour force in industry (%)	28.0[f]	21.9[e]
Labour force in agriculture (%)	14.8[e]	...
Agricultural production index (1979-1981=100)	126[d]	133[c]
Food production index (1979-1981=100)	127[d]	136[c]
Commercial energy production (1,000 Mt coal equiv)	2[f]	21[e]
Motor vehicles (per 1,000 population)	3	5[e]
Telephone lines (per 100 inhabitants)	0.1[h]	0.2[e]

Largest export industries	Major trading partners	1992
(% of exports)	(% of exports)	(% of imports)
...	... Belgium[i]	15
...	... France[j]	11
...	... U. R. Tanzania	9

Social indicators	1990/95
Growth rate of population (% per annum)	2.9
Age group 0-14 years (%)	46.3
Age group 60+ years (women and men, %)	2.6/1.8
Life expectancy at birth (women and men, years)	50/46
Infant mortality rate (per 1,000 births)	106
Total fertility rate (births per woman)	7
Contraceptive use (% of currently married women)	9[h]
Urban population (%)	6
Urban population growth rate (% per annum)	5.4
Rural population growth rate (% per annum)	2.7
Foreign-born (1985,%)	7.0
Refugees	271700
Government education expenditure (% of GDP)	4
Primary-secondary gross enrolment ratio (f and m, per 100)	35/44[k]
Third-level students (per 100,000 population)	66
Newspaper circulation (per 1,000 population)	4
Television receivers (per 1,000 population)	1
Parliamentary seats (women and men, %)	10/90

Environment	1990/91
Threatened species	15
Forested area (%)	2.4
CO_2 emissions (10,000 Mt)	60
Energy consumption (1,000 Mt coal equiv)	22
Precipitation (mm)	838[h]
Average temperature (January and July, centigrade)	23.1/22.1[f]

a Capital city. b December 1993. c 1992. d 1988. e 1991. f 1980.
g 1990. h 1987. i Includes Luxemburg. j Includes Monaco. k 1989.

Cambodia

Region South-eastern Asia
Location (longitude, latitude) 11°54'N 104°88'E
Currency riel
Population (1995 est., in thousands) 9447
Surface area (square kms) 181035
Population density (pop. per square km) 47
Sex ratio (females per 100 males) 108
Largest city (pop. in thousands) Phnom-Penh (968)
UN membership date 14-Dec-1955
Major language(s) Khmer, Chinese, Vietnamese

Economic indicators	1985	1994
Tourist arrivals (in thousands)	20[a]	88[b]
GDP (million US$)	650	825[c]
GDP (per capita US$)	89	96[c]
Long-term rate of change in GDP (% per annum)	−4.1	1.0[c]
Economically active female population (%)	57.9[d]	50.1
Economically active male population (%)	83.3[d]	84.6
Annual growth of econ. active female pop. (%)	1.0[e]	...
Annual growth of econ. active male pop. (%)	1.5[e]	...
Agricultural production index (1979-1981=100)	187[a]	188[b]
Food production index (1979-1981=100)	184[a]	182[b]
Commercial energy production (1,000 Mt coal equiv)	6[d]	4[c]
Telephone lines (per 100 inhabitants)	0.1[f]	0.1[c]

Social indicators	1990/95
Growth rate of population (% per annum)	2.5
Age group 0-14 years (%)	41.9
Age group 60+ years (women and men, %)	2.8/1.7
Life expectancy at birth (women and men, years)	52/50
Infant mortality rate (per 1,000 births)	116
Total fertility rate (births per woman)	5
Urban population (%)	13
Urban population growth rate (% per annum)	4.5
Rural population growth rate (% per annum)	2.2
Foreign-born (1985,%)	0.3
Television receivers (per 1,000 population)	8
Parliamentary seats (women and men, %)	4/96

Environment	1990/91
Threatened species	61
Forested area (%)	73.9
CO_2 emissions (10,000 Mt)	126
Energy consumption (1,000 Mt coal equiv)	26
Precipitation (mm)	1698[g]
Average temperature (January and July, centigrade)	25.6/27.1[d]

a 1988. b 1992. c 1991. d 1980. e 1990. f 1987. g Kompong-Cham.

Cameroon

Region	Middle Africa
Location (longitude, latitude)	4°01'N 9°43'E
Currency	CFA franc
Population (1995 est., in thousands)	13275
Surface area (square kms)	475442
Population density (pop. per square km)	26
Sex ratio (females per 100 males)	101
Largest city (pop. in thousands)	Douala (1001)
UN membership date	20-Sep-1960
Major language(s)	English, French, Beti, Ful

Economic indicators	1985	1994
Exchange rate (US$)	378.05	528.15
Consumer price index (1980=100)	181	...
Balance of payments, current account (million US$)	−562	−323[a]
Tourist arrivals (in thousands)	100[b]	62[c]
GDP (million US$)	8545	12788[a]
GDP (per capita US$)	857	1079[a]
Long-term rate of change in GDP (% per annum)	8.9	−1.9[a]
Gross fixed capital formation (% of GDP)	22.4	...
Economically active female population (%)	45.6[d]	39.3
Economically active male population (%)	88.8[d]	86.2
Annual growth of econ. active female pop. (%)	1.3[e]	...
Annual growth of econ. active male pop. (%)	2.2[e]	...
Agricultural production index (1979-1981=100)	113[b]	109[c]
Food production index (1979-1981=100)	110[b]	111[c]
Commercial energy production (1,000 Mt coal equiv)	4024[d]	11338[a]
Motor vehicles (per 1,000 population)	12	8[a]
Telephone lines (per 100 inhabitants)	0.3[f]	0.3[a]

Social indicators	1990/95
Growth rate of population (% per annum)	2.8
Age group 0-14 years (%)	44.0
Age group 60+ years (women and men, %)	3.0/2.5
Life expectancy at birth (women and men, years)	58/55
Infant mortality rate (per 1,000 births)	63
Total fertility rate (births per woman)	6
Contraceptive use (% of currently married women)	16
Urban population (%)	45
Urban population growth rate (% per annum)	5.0
Rural population growth rate (% per annum)	1.2
Foreign-born (1985, %)	2.4
Refugees	42200
Government education expenditure (% of GDP)	3
Primary-secondary gross enrolment ratio (f and m, per 100)	59/72[g]
Third-level students (per 100,000 population)	242[b]
Newspaper circulation (per 1,000 population)	7
Television receivers (per 1,000 population)	24
Parliamentary seats (women and men, %)	12/88

Environment	1990/91
Threatened species	76
Forested area (%)	51.6
CO_2 emissions (10,000 Mt)	525
Energy consumption (1,000 Mt coal equiv)	101
Precipitation (mm)	119[h]
Average temperature (January and July, centigrade)	24.1/22.2[d]

a 1991. b 1988. c 1992. d 1980. e 1990. f 1987. g 1989. h Yaoundé.

Canada

Region Northern America
Location (longitude, latitude) 43°66'N 79°41'W
Currency Canadian dollar
Population (1995 est., in thousands) 28537
Surface area (square kms) 9976139
Population density (pop. per square km) 3
Sex ratio (females per 100 males) 103
Largest city (pop. in thousands) Toronto (3463)
UN membership date 09-Nov-1945
Major language(s) English, French

Economic indicators	1985	1994
Exchange rate (US$)	1.40	1.34
Consumer price index (1980=100)	143	195[a]
Industrial production index (1980 = 100)	115	136[b]
Unemployment (%)	...	9.1[c]
Balance of payments, current account (million US$)	−2279	−19601[d]
Tourist arrivals (in thousands)	15485[e]	14741[f]
GDP (million US$)	347374	582010[g]
GDP (per capita US$)	13795	21562[g]
Long-term rate of change in GDP (% per annum)	4.7	−1.8[g]
Gross fixed capital formation (% of GDP)	19.9	19.9[g]
Economically active female population (%)	49.9[h]	49.4
Economically active male population (%)	77.9[h]	77.9
Annual growth of econ. active female pop. (%)	3.2[i]	...
Annual growth of econ. active male pop. (%)	1.6[i]	...
Labour force in industry (%)	28.5[h]	22.7[f]
Labour force in agriculture (%)	5.5[h]	4.3[f]
Agricultural production index (1979-1981=100)	102[e]	121[f]
Food production index (1979-1981=100)	103[e]	123[f]
Commercial energy production (1,000 Mt coal equiv)	288606[h]	404267[g]
Motor vehicles (per 1,000 population)	567	628[g]
Telephone lines (per 100 inhabitants)	25.3[j]	58.6[g]

Largest export industries	Major trading partners		1992
(% of exports)	(% of exports)	(% of imports)	
Metal manufacture 42	United States 78	United States	65
Mining, quarrying 11	Japan 5	Japan	7
Paper, paper products 10	United Kingdom 2	United Kingdom	3

Social indicators	1990/95
Growth rate of population (% per annum)	1.4
Age group 0-14 years (%)	20.7
Age group 60+ years (women and men, %)	9.1/7.0
Life expectancy at birth (women and men, years)	81/74
Infant mortality rate (per 1,000 births)	7
Total fertility rate (births per woman)	2
Urban population (%)	78
Urban population growth rate (% per annum)	1.6
Rural population growth rate (% per annum)	0.5
Foreign-born (1985,%)	15.5
Refugees	568200
Government education expenditure (% of GDP)	7
Primary-secondary gross enrolment ratio (f and m, per 100)	106/106
Third-level students (per 100,000 population)	5125
Newspaper circulation (per 1,000 population)	228[k]
Television receivers (per 1,000 population)	639
Intentional homicides (1986, per 100,000 population)	2
Parliamentary seats (women and men, %)	13/87

Environment	1990/91
Threatened species	47
Forested area (%)	36.0
CO_2 emissions (10,000 Mt)	112071
Energy consumption (1,000 Mt coal equiv)	10903
Precipitation (mm)	760[j]
Average temperature (January and July, centigrade)	−5.8/21.1[h]

a August 1994. b June 1994. c Labour force sample surveys. d 1993.
e 1988. f 1992. g 1991. h 1980. i 1990. j 1987. k 1989.

Cape Verde

Region	Southern Africa
Location (longitude, latitude)	15°00'N 23°70'W[a]
Currency	Cape Verdean escudo
Population (1995 est., in thousands)	419
Surface area (square kms)	4033
Population density (pop. per square km)	95
Sex ratio (females per 100 males)	112
Largest city (pop. in thousands)	Praia (62)[a]
UN membership date	16-Sep-1975
Major language(s)	Portuguese, Mandyak

Economic indicators	1985	1994
Exchange rate (US$)	85.38	82.88
Consumer price index (1980=100)	118	141[b]
Balance of payments, current account (million US$)	−9	−4[c]
GDP (million US$)	143	341[d]
GDP (per capita US$)	441	913[d]
Long-term rate of change in GDP (% per annum)	8.5	5.3[d]
Gross fixed capital formation (% of GDP)	45.5	...
Economically active female population (%)	29.4[e]	32.8
Economically active male population (%)	89.9[e]	90.0
Annual growth of econ. active female pop. (%)	3.3[f]	...
Annual growth of econ. active male pop. (%)	2.0[f]	...
Agricultural production index (1979-1981=100)	204[g]	170[c]
Food production index (1979-1981=100)	206[g]	172[c]
Motor vehicles (per 1,000 population)	9	...
Telephone lines (per 100 inhabitants)	1.7[h]	2.3[d]

Social indicators	1990/95
Growth rate of population (% per annum)	2.9
Age group 0-14 years (%)	43.0
Age group 60+ years (women and men, %)	3.6/2.4
Life expectancy at birth (women and men, years)	69/67
Infant mortality rate (per 1,000 births)	40
Total fertility rate (births per woman)	4
Urban population (%)	32
Urban population growth rate (% per annum)	5.0
Rural population growth rate (% per annum)	2.0
Foreign-born (1985,%)	1.1
Government education expenditure (% of GDP)	4
Primary-secondary gross enrolment ratio (f and m, per 100)	75/80[i]
Intentional homicides (1986, per 100,000 population)	2[e,j]
Parliamentary seats (women and men, %)	8/92

Environment	1990/91
Threatened species	7
Forested area (%)	0.2
CO_2 emissions (10,000 Mt)	23
Energy consumption (1,000 Mt coal equiv)	110

a Capital city. b December 1993. c 1992. d 1991. e 1980. f 1990.
g 1988. h 1987. i 1989. j Source: DYB92.

Central African Republic

Region	Middle Africa
Location (longitude, latitude)	4°23'N 18°34'E[a]
Currency	CFA franc
Population (1995 est., in thousands)	3429
Surface area (square kms)	622984
Population density (pop. per square km)	5
Sex ratio (females per 100 males)	106
Largest city (pop. in thousands)	Bangui (706)[a]
UN membership date	20-Sep-1960
Major language(s)	French, Banda, Gbaya, Sango, Sara

Economic indicators	1985	1994
Exchange rate (US$)	378.05	528.15
Consumer price index (1980=100)	147	138[b]
Balance of payments, current account (million US$)	−49	−57[c]
Tourist arrivals (in thousands)	5[d]	6[c]
GDP (million US$)	865	1443[e]
GDP (per capita US$)	329	467[e]
Long-term rate of change in GDP (% per annum)	3.3	−0.2[e]
Gross fixed capital formation (% of GDP)	12.4	...
Economically active female population (%)	74.1[f]	65.1
Economically active male population (%)	89.7[f]	87.5
Annual growth of econ. active female pop. (%)	1.1[g]	...
Annual growth of econ. active male pop. (%)	1.8[g]	...
Labour force in industry (%)	24.1[f]	46.9[g]
Labour force in agriculture (%)	32.4[f]	17.7[g]
Agricultural production index (1979-1981=100)	122[d]	128[c]
Food production index (1979-1981=100)	121[d]	129[c]
Commercial energy production (1,000 Mt coal equiv)	8[f]	10[e]
Motor vehicles (per 1,000 population)	19	5[g]
Telephone lines (per 100 inhabitants)	0.1[h]	0.2[e]

Social indicators	1990/95
Growth rate of population (% per annum)	2.6
Age group 0-14 years (%)	45.2
Age group 60+ years (women and men, %)	3.3/2.5
Life expectancy at birth (women and men, years)	49/45
Infant mortality rate (per 1,000 births)	105
Total fertility rate (births per woman)	6
Urban population (%)	51
Urban population growth rate (% per annum)	4.3
Rural population growth rate (% per annum)	1.0
Foreign-born (1985, %)	3.8
Refugees	19000
Government education expenditure (% of GDP)	3
Primary-secondary gross enrolment ratio (f and m, per 100)	30/52[i]
Third-level students (per 100,000 population)	118[i]
Newspaper circulation (per 1,000 population)	1
Television receivers (per 1,000 population)	4
Parliamentary seats (women and men, %)	4/96[h]

Environment	1990/91
Threatened species	17
Forested area (%)	57.5
CO_2 emissions (10,000 Mt)	57
Energy consumption (1,000 Mt coal equiv)	35
Precipitation (mm)	1560[h]
Average temperature (January and July, centigrade)	25.9/25.1[f]

a Capital city. b March 1994. c 1992. d 1988. e 1991. f 1980. g 1990. h 1987. i 1989.

Chad

Region	Middle Africa
Location (longitude, latitude)	12°07'N 15°03'E[a]
Currency	CFA franc
Population (1995 est., in thousands)	6361
Surface area (square kms)	1284000
Population density (pop. per square km)	5
Sex ratio (females per 100 males)	103
Largest city (pop. in thousands)	N'Djamena (728)[a]
UN membership date	20-Sep-1960
Major language(s)	French, Arabic, Sara, Tubu

Economic indicators	1985	1994
Exchange rate (US$)	378.05	528.15
Consumer price index (1980=100)	129	96[b]
Balance of payments, current account (million US$)	–87	–80[c]
Tourist arrivals (in thousands)	21[d]	20[e]
GDP (million US$)	780	1290[c]
GDP (per capita US$)	155	227[c]
Long-term rate of change in GDP (% per annum)	21.9	–1.7[c]
Economically active female population (%)	25.3[f]	22.5
Economically active male population (%)	90.5[f]	89.2
Annual growth of econ. active female pop. (%)	1.4[b]	...
Annual growth of econ. active male pop. (%)	2.0[b]	...
Labour force in industry (%)	23.6[g]	52.0[c]
Labour force in agriculture (%)	19.1[g]	11.2[c]
Agricultural production index (1979-1981=100)	132[d]	141[e]
Food production index (1979-1981=100)	130[d]	140[e]
Motor vehicles (per 1,000 population)	3	...
Telephone lines (per 100 inhabitants)	0.1[h]	0.1[c]

Social indicators	1990/95
Growth rate of population (% per annum)	2.7
Age group 0-14 years (%)	43.4
Age group 60+ years (women and men, %)	3.1/2.5
Life expectancy at birth (women and men, years)	49/46
Infant mortality rate (per 1,000 births)	122
Total fertility rate (births per woman)	6
Urban population (%)	37
Urban population growth rate (% per annum)	5.9
Rural population growth rate (% per annum)	1.1
Foreign-born (1985,%)	0.3
Government education expenditure (% of GDP)	2
Primary-secondary gross enrolment ratio (f and m, per 100)	19/46[i]
Third-level students (per 100,000 population)	34[j]
Newspaper circulation (per 1,000 population)	0
Television receivers (per 1,000 population)	1

Environment	1990/91
Threatened species	20
Forested area (%)	9.9
CO_2 emissions (10,000 Mt)	69
Energy consumption (1,000 Mt coal equiv)	21
Precipitation (mm)	646[k]
Average temperature (January and July, centigrade)	23.5/27.5[f]

a Capital city. b 1990. c 1991. d 1988. e 1992. f 1980. g 1986. h 1987. i 1989. j 1984. k Fort-Lamy.

Chile

Region	South America
Location (longitude, latitude)	33°44'S 70°67'W
Currency	Chilean peso
Population (1995 est., in thousands)	14237
Surface area (square kms)	756945
Population density (pop. per square km)	18
Sex ratio (females per 100 males)	102
Largest city (pop. in thousands)	Santiago (4870)
UN membership date	24-Oct-1945
Major language(s)	Spanish

Economic indicators	1985	1994
Exchange rate (US$)	183.86	412.32
Consumer price index (1980=100)	262	1145
Unemployment (%)	...	6.1[a]
Balance of payments, current account (million US$)	−1413	−583[b]
Tourist arrivals (in thousands)	624[c]	1283[b]
GDP (million US$)	15996	31311[d]
GDP (per capita US$)	1320	2339[d]
Long-term rate of change in GDP (% per annum)	2.0	6.0[d]
Gross fixed capital formation (% of GDP)	14.2	17.8[d]
Economically active female population (%)	26.3[e]	29.1
Economically active male population (%)	73.0[e]	74.8
Annual growth of econ. active female pop. (%)	3.7[f]	...
Annual growth of econ. active male pop. (%)	2.0[f]	...
Labour force in industry (%)	23.7[e]	26.5[b]
Labour force in agriculture (%)	16.3[e]	18.0[b]
Agricultural production index (1979-1981=100)	120[c]	141[b]
Food production index (1979-1981=100)	121[c]	142[b]
Commercial energy production (1,000 Mt coal equiv)	5722[e]	7225[d]
Motor vehicles (per 1,000 population)	73	88[d]
Telephone lines (per 100 inhabitants)	4.6[g]	7.4[d]

Social indicators	1990/95
Growth rate of population (% per annum)	1.6
Age group 0-14 years (%)	30.5
Age group 60+ years (women and men, %)	5.3/3.9
Life expectancy at birth (women and men, years)	76/69
Infant mortality rate (per 1,000 births)	17
Total fertility rate (births per woman)	3
Urban population (%)	86
Urban population growth rate (% per annum)	1.9
Rural population growth rate (% per annum)	−0.2
Foreign-born (1985,%)	0.7
Refugees	100
Government education expenditure (% of GDP)	3
Primary-secondary gross enrolment ratio (f and m, per 100)	91/90
Third-level students (per 100,000 population)	1629[h]
Newspaper circulation (per 1,000 population)	455
Television receivers (per 1,000 population)	209
Intentional homicides (1986, per 100,000 population)	14
Parliamentary seats (women and men, %)	6/94

Environment	1990/91
Threatened species	106
Forested area (%)	11.6
CO_2 emissions (10,000 Mt)	8877
Energy consumption (1,000 Mt coal equiv)	1182

a Labour force sample surveys. b 1992. c 1988. d 1991. e 1980. f 1990. g 1987. h 1985.

China

Region	Eastern Asia
Location (longitude, latitude)	31°23'N 121°36'E
Currency	yuan renminbi
Population (1995 est., in thousands)	1238319
Surface area (square kms)	9596961
Population density (pop. per square km)	120
Sex ratio (females per 100 males)	95
Largest city (pop. in thousands)	Shanghai (13447)
UN membership date	24-Oct-1945
Major language(s)	Chinese

Economic indicators	1985	1994
Exchange rate (US$)	3.20	8.53
Consumer price index (1980=100)	123	203[a]
Balance of payments, current account (million US$)	−11417	6401[b]
Tourist arrivals (in thousands)	12361[c]	16512[b]
GDP (million US$)	290360	370854[d]
GDP (per capita US$)	276	322[d]
Long-term rate of change in GDP (% per annum)	12.1	6.6
Economically active female population (%)	69.5[e]	70.1
Economically active male population (%)	87.9[e]	87.1
Annual growth of econ. active female pop. (%)	2.8[a]	...
Annual growth of econ. active male pop. (%)	2.4[a]	...
Labour force in industry (%)	48.9[f]	48.0[g]
Labour force in agriculture (%)	8.7[f]	7.9[g]
Agricultural production index (1979-1981=100)	143[c]	169[b]
Food production index (1979-1981=100)	140[c]	168[b]
Commercial energy production (1,000 Mt coal equiv)	612383[e]	1014037[b]
Telephone lines (per 100 inhabitants)	0.4[h]	0.7[d]

Largest export industries		Major trading partners			1992
	(% of exports)	(% of exports)		(% of imports)	
Textiles	36	Hong Kong	44	Hong Kong	26
Metal manufacture	23	Japan	14	Japan	17
Chemicals	10	United States	10	United States	11

Social indicators	1990/95
Growth rate of population (% per annum)	1.4
Age group 0-14 years (%)	27.3
Age group 60+ years (women and men, %)	4.9/4.6
Life expectancy at birth (women and men, years)	73/69
Infant mortality rate (per 1,000 births)	27
Total fertility rate (births per woman)	2
Contraceptive use (% of currently married women)	83
Urban population (%)	30
Urban population growth rate (% per annum)	4.3
Rural population growth rate (% per annum)	0.3
Foreign-born (1985,%)	0.3
Refugees	288100
Government education expenditure (% of GDP)	2
Primary-secondary gross enrolment ratio (f and m, per 100)	81/93
Third-level students (per 100,000 population)	188
Newspaper circulation (per 1,000 population)	36[e]
Television receivers (per 1,000 population)	31
Intentional homicides (1986, per 100,000 population)	1
Parliamentary seats (women and men, %)	21/79

Environment	1990/91
Threatened species	194
Forested area (%)	13.2
CO_2 emissions (10,000 Mt)	694154
Energy consumption (1,000 Mt coal equiv)	811
Precipitation (mm)	2100[h]
Average temperature (January and July, centigrade)	15.2/28.4[e]

a 1990. b 1992. c 1988. d 1991. e 1980. f 1985. g 1989. h 1987.

Colombia

Region	South America
Location (longitude, latitude)	4°59'N 74°09'W
Currency	Colombian peso
Population (1995 est., in thousands)	35101
Surface area (square kms)	1138914
Population density (pop. per square km)	30
Sex ratio (females per 100 males)	102
Largest city (pop. in thousands)	Bogota (4851)
UN membership date	05-Nov-1945
Major language(s)	Spanish

Economic indicators	1985	1994
Exchange rate (US$)	172.20[a]	842.00[a]
Consumer price index (1980=100)	277	1911[b]
Industrial production index (1980 = 100)	114	153
Unemployment (%)	...	9.9[c]
Balance of payments, current account (million US$)	−1809	912[d]
Tourist arrivals (in thousands)	829[e]	1076[d]
GDP (million US$)	34896	41450[f]
GDP (per capita US$)	1184	1261[f]
Long-term rate of change in GDP (% per annum)	3.1	2.1[f]
Gross fixed capital formation (% of GDP)	17.5	16.6[g]
Economically active female population (%)	22.1[h]	22.4
Economically active male population (%)	77.4[h]	78.7
Annual growth of econ. active female pop. (%)	3.0[g]	...
Annual growth of econ. active male pop. (%)	2.7[g]	...
Labour force in industry (%)	34.0[h]	30.9[d]
Labour force in agriculture (%)	1.4[h]	1.4
Agricultural production index (1979-1981=100)	119[e]	138[d]
Food production index (1979-1981=100)	125[e]	139[d]
Commercial energy production (1,000 Mt coal equiv)	20158[h]	61663[f]
Motor vehicles (per 1,000 population)	41	...
Telephone lines (per 100 inhabitants)	6.3[i]	8.0[f]

Largest export industries		Major trading partners			1992
	(% of exports)	(% of exports)		(% of imports)	
Agriculture	34	United States	39	United States	36
Mining, quarrying	27	Germany	9	Japan	9
Textiles	12	Venezuela	9	Venezuela	7

Social indicators	1990/95
Growth rate of population (% per annum)	1.7
Age group 0-14 years (%)	32.9
Age group 60+ years (women and men, %)	3.6/3.0
Life expectancy at birth (women and men, years)	72/66
Infant mortality rate (per 1,000 births)	37
Total fertility rate (births per woman)	3
Contraceptive use (% of currently married women)	66
Urban population (%)	73
Urban population growth rate (% per annum)	2.4
Rural population growth rate (% per annum)	−0.3
Foreign-born (1985, %)	0.3
Refugees	500
Government education expenditure (% of GDP)	3[j]
Primary-secondary gross enrolment ratio (f and m, per 100)	82/76
Third-level students (per 100,000 population)	1466[j]
Newspaper circulation (per 1,000 population)	62
Television receivers (per 1,000 population)	116
Intentional homicides (1986, per 100,000 population)	2
Parliamentary seats (women and men, %)	4/96

Environment	1990/91
Threatened species	106
Forested area (%)	44.2
CO_2 emissions (10,000 Mt)	15694
Energy consumption (1,000 Mt coal equiv)	838

a Selling rate. b January 1994. c Labour force sample surveys. d 1992.
e 1988. f 1991. g 1990. h 1980. i 1987. j 1989.

Comoros

Region Eastern Africa
Location (longitude, latitude) 11°42'S 43°14'E[a]
Currency Comoros franc
Population (1995 est., in thousands) 653
Surface area (square kms) 2235
Population density (pop. per square km) 255
Sex ratio (females per 100 males) 97
Largest city (pop. in thousands) Moroni (24)[a]
UN membership date 12-Nov-1975
Major language(s) French

Economic indicators	1985	1994
Exchange rate (US$)	378.05	396.11
Balance of payments, current account (million US$)	–14	–9[b]
Tourist arrivals (in thousands)	8[c]	19[d]
GDP (million US$)	114	245[b]
GDP (per capita US$)	252	435[b]
Long-term rate of change in GDP (% per annum)	2.7	1.8[b]
Economically active female population (%)	64.1[e]	56.7
Economically active male population (%)	93.3[e]	90.7
Annual growth of econ. active female pop. (%)	2.6[f]	...
Annual growth of econ. active male pop. (%)	3.1[f]	...
Agricultural production index (1979-1981=100)	120[c]	128[d]
Food production index (1979-1981=100)	120[c]	128[d]
Commercial energy production (1,000 Mt coal equiv)	0[e]	0[b]
Motor vehicles (per 1,000 population)	11	...
Telephone lines (per 100 inhabitants)	0.4[g]	0.7[b]

Social indicators	1990/95
Growth rate of population (% per annum)	3.7
Age group 0-14 years (%)	48.7
Age group 60+ years (women and men, %)	2.1/1.8
Life expectancy at birth (women and men, years)	57/56
Infant mortality rate (per 1,000 births)	89
Total fertility rate (births per woman)	7
Urban population (%)	31
Urban population growth rate (% per annum)	5.7
Rural population growth rate (% per annum)	2.9
Foreign-born (1985,%)	3.3
Government education expenditure (% of GDP)	4[h]
Primary-secondary gross enrolment ratio (f and m, per 100)	46/58[h]
Parliamentary seats (women and men, %)	0/100[g]

Environment	1990/91
Threatened species	16
Forested area (%)	15.7
CO_2 emissions (10,000 Mt)	18
Energy consumption (1,000 Mt coal equiv)	55
Precipitation (mm)	2639[g]
Average temperature (January and July, centigrade)	26.2/22.8[e]

a Capital city. b 1991. c 1988. d 1992. e 1980. f 1990. g 1987. h 1986.

Congo

Region	Middle Africa
Location (longitude, latitude)	4°50'S 15°14'E[a]
Currency	CFA franc
Population (1995 est., in thousands)	2590
Surface area (square kms)	342000
Population density (pop. per square km)	7
Sex ratio (females per 100 males)	104
Largest city (pop. in thousands)	Brazzaville (619)[a]
UN membership date	20-Sep-1960
Major language(s)	French, Congo, Teke, Mboshi

Economic indicators	1985	1994
Exchange rate (US$)	378.05	528.15
Consumer price index (1980=100)	170	204
Balance of payments, current account (million US$)	−161	−308[b]
Tourist arrivals (in thousands)	39[c]	37[b]
GDP (million US$)	2161	2909[d]
GDP (per capita US$)	1124	1266[d]
Long-term rate of change in GDP (% per annum)	−1.2	−2.4[d]
Gross fixed capital formation (% of GDP)	28.5	...
Economically active female population (%)	53.4[e]	50.5
Economically active male population (%)	85.5[e]	82.5
Annual growth of econ. active female pop. (%)	2.0[f]	...
Annual growth of econ. active male pop. (%)	2.2[f]	...
Agricultural production index (1979-1981=100)	123[c]	112[b]
Food production index (1979-1981=100)	123[c]	112[b]
Commercial energy production (1,000 Mt coal equiv)	4720[e]	11283[d]
Motor vehicles (per 1,000 population)	24	21[d]
Telephone lines (per 100 inhabitants)	0.5[g]	0.7[d]

Social indicators	1990/95
Growth rate of population (% per annum)	3.0
Age group 0-14 years (%)	45.7
Age group 60+ years (women and men, %)	2.8/2.3
Life expectancy at birth (women and men, years)	54/49
Infant mortality rate (per 1,000 births)	82
Total fertility rate (births per woman)	6
Urban population (%)	43
Urban population growth rate (% per annum)	4.4
Rural population growth rate (% per annum)	2.0
Foreign-born (1985,%)	5.2
Refugees	9500
Government education expenditure (% of GDP)	6
Third-level students (per 100,000 population)	470
Newspaper circulation (per 1,000 population)	8
Television receivers (per 1,000 population)	6
Parliamentary seats (women and men, %)	10/90[g]

Environment	1990/91
Threatened species	24
Forested area (%)	61.9
CO_2 emissions (10,000 Mt)	550
Energy consumption (1,000 Mt coal equiv)	379
Precipitation (mm)	1371[g]
Average temperature (January and July, centigrade)	25.6/21.7[e]

a Capital city. b 1992. c 1988. d 1991. e 1980. f 1990. g 1987.

Cook Islands

Region Oceania-Polynesia
Location (longitude, latitude) 21°21'S 159°46'W[a]
Currency New Zealand dollar
Surface area (square kms) 236
Population density (pop. per square km) 76
Largest city (pop. in thousands) Avarua (4)[a]
Major language(s) English

Economic indicators	1985	1994
Consumer price index (1980=100)	184	115
Tourist arrivals (in thousands)	34[b]	50[c]
GDP (million US$)	26	55[d]
GDP (per capita US$)	1438	3255[d]
Long-term rate of change in GDP (% per annum)	8.8	3.9[d]
Telephone lines (per 100 inhabitants)	17.6[f]	...

Largest export industries	Major trading partners		1992
(% of exports)	(% of exports)		(% of imports)
...	New Zealand	40	...
...	Hong Kong	25	...
...	Japan	22	...

Social indicators	1990/95
Contraceptive use (% of currently married women)	38[g]
Urban population (%)	26
Urban population growth rate (% per annum)	0.7
Rural population growth rate (% per annum)	-0.4
Foreign-born (1985,%)	16.4
Newspaper circulation (per 1,000 population)	118
Television receivers (per 1,000 population)	182

Environment	1990/91
Threatened species	8
CO_2 emissions (10,000 Mt)	6
Energy consumption (1,000 Mt coal equiv)	588

a Capital city. b 1988. c 1992. d 1991. e 1980. f 1990. g 1989.

Costa Rica

Region Central America
Location (longitude, latitude) 9°93'N 84°07'W
Currency Costa Rican colone
Population (1995 est., in thousands) 3424
Surface area (square kms) 51100
Population density (pop. per square km) 60
Sex ratio (females per 100 males) 98
Largest city (pop. in thousands) San Jose (760)
UN membership date 02-Nov-1945
Major language(s) Spanish

Economic indicators	1985	1994
Exchange rate (US$)	53.70	160.05
Consumer price index (1980=100)	443	1881[a]
Industrial production index (1980 = 100)	104	138[b]
Balance of payments, current account (million US$)	−291	−371[c]
Tourist arrivals (in thousands)	329[d]	610[c]
GDP (million US$)	3923	5523[e]
GDP (per capita US$)	1485	1774[e]
Long-term rate of change in GDP (% per annum)	0.7	1.0[e]
Gross fixed capital formation (% of GDP)	19.3	22.5[f]
Economically active female population (%)	23.2[g]	23.8
Economically active male population (%)	84.0[g]	83.3
Annual growth of econ. active female pop. (%)	4.6[f]	...
Annual growth of econ. active male pop. (%)	3.3[f]	...
Labour force in industry (%)	24.0[g]	26.2[c]
Labour force in agriculture (%)	27.4[g]	24.1[c]
Agricultural production index (1979-1981=100)	120[d]	145[c]
Food production index (1979-1981=100)	116[d]	143[c]
Commercial energy production (1,000 Mt coal equiv)	259[g]	448[e]
Motor vehicles (per 1,000 population)	69	91[e]
Telephone lines (per 100 inhabitants)	8.6[h]	9.8[e]

Social indicators	1990/95
Growth rate of population (% per annum)	2.4
Age group 0-14 years (%)	35.0
Age group 60+ years (women and men, %)	3.7/3.3
Life expectancy at birth (women and men, years)	79/74
Infant mortality rate (per 1,000 births)	14
Total fertility rate (births per woman)	3
Contraceptive use (% of currently married women)	70[i]
Urban population (%)	50
Urban population growth rate (% per annum)	3.5
Rural population growth rate (% per annum)	1.4
Foreign-born (1985,%)	4.3
Refugees	114400
Government education expenditure (% of GDP)	4
Primary-secondary gross enrolment ratio (f and m, per 100)	77/77
Third-level students (per 100,000 population)	2477
Newspaper circulation (per 1,000 population)	101
Television receivers (per 1,000 population)	140
Intentional homicides (1986, per 100,000 population)	4
Parliamentary seats (women and men, %)	12/88

Environment	1990/91
Threatened species	37
Forested area (%)	32.1
CO_2 emissions (10,000 Mt)	887
Energy consumption (1,000 Mt coal equiv)	559
Precipitation (mm)	1944[h]
Average temperature (January and July, centigrade)	19.0/20.6[g]

a July 1994. b April 1994. c 1992. d 1988. e 1991. f 1990. g 1980.
h 1987. i 1986.

Côte d'Ivoire

Region	Southern Africa
Location (longitude, latitude)	5°32'N 4°04'W
Currency	CFA franc
Population (1995 est., in thousands)	14401
Surface area (square kms)	322463
Population density (pop. per square km)	39
Sex ratio (females per 100 males)	97
Largest city (pop. in thousands)	Abidjan (2168)
UN membership date	20-Sep-1960
Major language(s)	French, Akan, Senufo, Bete, Dan

Economic indicators	1985	1994
Exchange rate (US$)	378.05	528.15
Consumer price index (1980=100)	131	165[a]
Industrial production index (1980 = 100)	112	105[b]
Balance of payments, current account (million US$)	64	−1229[c]
Tourist arrivals (in thousands)	178[d]	217[b]
GDP (million US$)	6978	12360[e]
GDP (per capita US$)	702	994[e]
Long-term rate of change in GDP (% per annum)	5.8	1.7[e]
Gross fixed capital formation (% of GDP)	11.8	...
Economically active female population (%)	52.1[f]	46.8
Economically active male population (%)	89.9[f]	86.8
Annual growth of econ. active female pop. (%)	2.2[a]	...
Annual growth of econ. active male pop. (%)	3.0[a]	...
Labour force in industry (%)	26.3[f]	20.1[a]
Labour force in agriculture (%)	15.1[f]	13.8[a]
Agricultural production index (1979-1981=100)	133[d]	134[b]
Food production index (1979-1981=100)	141[d]	147[b]
Commercial energy production (1,000 Mt coal equiv)	338[f]	653[e]
Motor vehicles (per 1,000 population)	28	20[a]
Telephone lines (per 100 inhabitants)	0.6[g]	0.6[e]

Social indicators	1990/95
Growth rate of population (% per annum)	3.7
Age group 0-14 years (%)	49.0
Age group 60+ years (women and men, %)	2.1/2.2
Life expectancy at birth (women and men, years)	53/50
Infant mortality rate (per 1,000 births)	91
Total fertility rate (births per woman)	7
Urban population (%)	44
Urban population growth rate (% per annum)	5.2
Rural population growth rate (% per annum)	2.6
Foreign-born (1985,%)	26.1
Refugees	174100
Government education expenditure (% of GDP)	6[h]
Primary-secondary gross enrolment ratio (f and m, per 100)	38/58
Third-level students (per 100,000 population)	240[f]
Newspaper circulation (per 1,000 population)	8
Television receivers (per 1,000 population)	59
Parliamentary seats (women and men, %)	5/95

Environment	1990/91
Threatened species	38
Forested area (%)	22.9
CO_2 emissions (10,000 Mt)	1741
Energy consumption (1,000 Mt coal equiv)	217
Precipitation (mm)	2144[g]
Average temperature (January and July, centigrade)	26.7/25.1[f]

a 1990. b 1992. c 1993. d 1988. e 1991. f 1980. g 1987. h 1985.

Croatia

Region	Southern Europe
Location (longitude, latitude)	45°79'N 15°97'E
Currency	new dinar
Largest city (pop. in thousands)	Zagreb (704)[a]
UN membership date	22-May-1992
Major language(s)	Croatian

Economic indicators	1985	1994
Consumer price index (1980=100)	7	53974
GDP (million US$)	11162	34259[b]
GDP (per capita US$)	2397	7232[b]
Long-term rate of change in GDP (% per annum)	0.1	−29.9[b]
Economically active female population (%)	45.9[c]	48.5
Economically active male population (%)	73.4[c]	68.6
Labour force in industry (%)	48.1[d]	43.6[e]
Labour force in agriculture (%)	4.7[d]	5.3[e]
Telephone lines (per 100 inhabitants)	17.4[f]	18.8[b]

Largest export industries	Major trading partners			1992
(% of exports)	(% of exports)		(% of imports)	
...	Slovenia	24	Slovenia	20
...	Italy	20	Germany	17
...	Germany	17	Italy	17

Social indicators	1990/95
Refugees	648000
Television receivers (per 1,000 population)	221

a Capital city. b 1991. c 1980. d 1981. e 1992. f 1990.

Cuba

Region	Caribbean
Location (longitude, latitude)	23°12'N 82°38'W
Currency	Cuban peso
Population (1995 est., in thousands)	11091
Surface area (square kms)	110861
Population density (pop. per square km)	97
Sex ratio (females per 100 males)	99
Largest city (pop. in thousands)	Havana (2124)
UN membership date	24-Oct-1945
Major language(s)	Spanish

Economic indicators	1985	1994
Exchange rate (US$)	0.91	0.70
Tourist arrivals (in thousands)	298[a]	410[b]
GDP (million US$)	17968	18738[c]
GDP (per capita US$)	1783	1749[c]
Long-term rate of change in GDP (% per annum)	1.6	−15.0[c]
Economically active female population (%)	32.5[d]	37.6
Economically active male population (%)	73.0[d]	75.7
Annual growth of econ. active female pop. (%)	5.5[e]	...
Annual growth of econ. active male pop. (%)	1.8[e]	...
Labour force in industry (%)	32.5[d]	31.3[a]
Labour force in agriculture (%)	19.6[d]	19.2[a]
Agricultural production index (1979-1981=100)	110[a]	93[b]
Food production index (1979-1981=100)	110[a]	93[b]
Commercial energy production (1,000 Mt coal equiv)	426[d]	1194[c]
Motor vehicles (per 1,000 population)	38	...
Telephone lines (per 100 inhabitants)	3.6[f]	3.2[c]

Social indicators	1990/95
Growth rate of population (% per annum)	0.9
Age group 0-14 years (%)	23.2
Age group 60+ years (women and men, %)	6.4/6.0
Life expectancy at birth (women and men, years)	78/74
Infant mortality rate (per 1,000 births)	14
Total fertility rate (births per woman)	2
Contraceptive use (% of currently married women)	70[f]
Urban population (%)	76
Urban population growth rate (% per annum)	1.5
Rural population growth rate (% per annum)	−1.0
Foreign-born (1985,%)	0.8
Refugees	5100
Government education expenditure (% of GDP)	6[g]
Primary-secondary gross enrolment ratio (f and m, per 100)	97/93[g]
Third-level students (per 100,000 population)	2304[g]
Newspaper circulation (per 1,000 population)	172
Television receivers (per 1,000 population)	163
Parliamentary seats (women and men, %)	23/77

Environment	1990/91
Threatened species	38
Forested area (%)	24.9
CO_2 emissions (10,000 Mt)	9388
Energy consumption (1,000 Mt coal equiv)	1436

a 1988. b 1992. c 1991. d 1980. e 1990. f 1987. g 1989.

Cyprus

Region Western Asia
Location (longitude, latitude) 35°16'N 33°35'E[a]
Currency Cyprus pound
Population (1995 est., in thousands) 736
Surface area (square kms) 9251
Population density (pop. per square km) 77
Sex ratio (females per 100 males) 100
Largest city (pop. in thousands) Nicosia (162)[a]
UN membership date 20-Sep-1960
Major language(s) Greek, Turkish

Economic indicators	1985	1994
Exchange rate (US$)	0.54	0.47
Consumer price index (1980=100)	138	201
Industrial production index (1980 = 100)	116	163
Balance of payments, current account (million US$)	−161	−242[b]
Tourist arrivals (in thousands)	1112[c]	1991[b]
GDP (million US$)	2430	5743[d]
GDP (per capita US$)	3653	8101[d]
Long-term rate of change in GDP (% per annum)	4.7	1.0[d]
Gross fixed capital formation (% of GDP)	27.2	24.2[e]
Economically active female population (%)	42.1[f]	44.6
Economically active male population (%)	80.4[f]	80.0
Annual growth of econ. active female pop. (%)	1.5[e]	...
Annual growth of econ. active male pop. (%)	1.1[e]	...
Labour force in industry (%)	33.8[f]	28.4[d]
Labour force in agriculture (%)	19.2[f]	13.3[d]
Agricultural production index (1979-1981=100)	111[c]	115[b]
Food production index (1979-1981=100)	112[c]	116[b]
Motor vehicles (per 1,000 population)	260	390[d]
Telephone lines (per 100 inhabitants)	27.8[g]	39.2[d]

Largest export industries		Major trading partners		1992
	(% of exports)	(% of exports)		(% of imports)
Food, beverages, tobacco	27	United Kingdom 19	United Kingdom	11
Textiles	27	Lebanon 12	Japan	11
Metal manufacture	17	Greece 8	Italy	10

Social indicators	1990/95
Growth rate of population (% per annum)	0.9
Age group 0-14 years (%)	25.5
Age group 60+ years (women and men, %)	7.7/6.3
Life expectancy at birth (women and men, years)	79/75
Infant mortality rate (per 1,000 births)	9
Total fertility rate (births per woman)	2
Urban population (%)	56
Urban population growth rate (% per annum)	2.2
Rural population growth rate (% per annum)	−0.6
Foreign-born (1985,%)	1.4
Government education expenditure (% of GDP)	4
Primary-secondary gross enrolment ratio (f and m, per 100)	98/97
Third-level students (per 100,000 population)	935
Newspaper circulation (per 1,000 population)	110
Television receivers (per 1,000 population)	144
Intentional homicides (1986, per 100,000 population)	1
Parliamentary seats (women and men, %)	5/95

Environment	1990/91
Threatened species	10
Forested area (%)	13.3
CO_2 emissions (10,000 Mt)	1223
Energy consumption (1,000 Mt coal equiv)	2587
Precipitation (mm)	339[g]
Average temperature (January and July, centigrade)	10.6/28.1[f]

a Capital city. b 1992. c 1988. d 1991. e 1990. f 1980. g 1987.

Czech Republic

Region Eastern Europe
Location (longitude, latitude) 50°09'N 14°43'E
Currency koruny
Largest city (pop. in thousands) Prague (1212)[a]
UN membership date 19-Jan-1993
Major language(s) Czech

Economic indicators	1985	1994
Exchange rate (US$)	...	27.80
Consumer price index (1980=100)	...	147[b]
GDP (million US$)	27637	24308[c]
GDP (per capita US$)	2674	2359[c]
Long-term rate of change in GDP (% per annum)	0.6	−14.2[c]
Gross fixed capital formation (% of GDP)	25.6	23.1[c]
Economically active female population (%)	61.4[d]	61.9
Economically active male population (%)	75.4[d]	73.4

Social indicators	1990/95
Contraceptive use (% of currently married women)	78
Intentional homicides (1986, per 100,000 population)	1

Environment	1990/91
Threatened species	42
Precipitation (mm)	508[e]
Average temperature (January and July, centigrade)	−2.6/17.9[d]

a Capital city. b July 1994. c 1991. d 1980. e 1987.

Denmark

Region	Northern Europe
Location (longitude, latitude)	55°68'N 12°58'E
Currency	Danish kroner
Population (1995 est., in thousands)	5192
Surface area (square kms)	43077
Population density (pop. per square km)	120
Sex ratio (females per 100 males)	102
Largest city (pop. in thousands)	Copenhagen (1331)
UN membership date	24-Oct-1945
Major language(s)	Danish

Economic indicators	1985	1994
Exchange rate (US$)	8.97	6.07
Consumer price index (1980=100)	146	192
Industrial production index (1980 = 100)	121	132[a]
Unemployment (%)	...	12.0
Balance of payments, current account (million US$)	−2767	5347[b]
Tourist arrivals (in thousands)	1150[c]	1543[d]
GDP (million US$)	58048	130287[e]
GDP (per capita US$)	11333	25308[e]
Long-term rate of change in GDP (% per annum)	4.3	1.2[e]
Gross fixed capital formation (% of GDP)	18.7	16.9[e]
Economically active female population (%)	57.3[f]	58.9
Economically active male population (%)	76.6[f]	76.6
Annual growth of econ. active female pop. (%)	2.0[g]	...
Annual growth of econ. active male pop. (%)	0.2[g]	...
Labour force in industry (%)	29.4[h]	27.3[e]
Labour force in agriculture (%)	7.4[h]	5.6[e]
Agricultural production index (1979-1981=100)	122[c]	120[d]
Food production index (1979-1981=100)	122[c]	120[d]
Commercial energy production (1,000 Mt coal equiv)	431[f]	15390[e]
Motor vehicles (per 1,000 population)	348	373[e]
Telephone lines (per 100 inhabitants)	52.9[i]	57.7[e]

Largest export industries		Major trading partners			1992
	(% of exports)		(% of exports)		(% of imports)
Metal manufacture	38	Germany	22	Germany	23
Food, beverages, tobacco	21	Sweden	10	Sweden	11
Chemicals	13	United Kingdom	10	United Kingdom	8

Social indicators	1990/95
Growth rate of population (% per annum)	0.2
Age group 0-14 years (%)	17.1
Age group 60+ years (women and men, %)	11.4/8.6
Life expectancy at birth (women and men, years)	78/73
Infant mortality rate (per 1,000 births)	7
Total fertility rate (births per woman)	2
Contraceptive use (% of currently married women)	78[c]
Urban population (%)	85
Urban population growth rate (% per annum)	0.4
Rural population growth rate (% per annum)	−0.7
Foreign-born (1985, %)	3.5
Refugees	58300
Government education expenditure (% of GDP)	7[j]
Primary-secondary gross enrolment ratio (f and m, per 100)	104/102
Third-level students (per 100,000 population)	2466[c]
Newspaper circulation (per 1,000 population)	352
Television receivers (per 1,000 population)	536
Intentional homicides (1986, per 100,000 population)	6
Parliamentary seats (women and men, %)	33/67

Environment	1990/91
Threatened species	28
Forested area (%)	11.4
CO_2 emissions (10,000 Mt)	17209
Energy consumption (1,000 Mt coal equiv)	5084
Precipitation (mm)	602[i]
Average temperature (January and July, centigrade)	0.1/17.8[f]

a December 1993. b 1993. c 1988. d 1992. e 1991. f 1980. g 1990. h 1981. i 1987. j 1989.

Djibouti

Region	Eastern Africa
Location (longitude, latitude)	11°30'N 43°00'E[a]
Currency	Djibouti franc
Population (1995 est., in thousands)	511
Surface area (square kms)	23200
Population density (pop. per square km)	18
Sex ratio (females per 100 males)	99
Largest city (pop. in thousands)	Djibouti (355)[a]
UN membership date	20-Sep-1977
Major language(s)	Djibouti, French, Somali, Afar

Economic indicators	1985	1994
Exchange rate (US$)	177.72	177.72
Tourist arrivals (in thousands)	29[b]	40[c]
GDP (million US$)	411	596[d]
GDP (per capita US$)	1082	1315[d]
Long-term rate of change in GDP (% per annum)	0.6	2.7[d]
Economically active female population (%)	65.9[e]	57.4
Economically active male population (%)	91.4[e]	89.5
Motor vehicles (per 1,000 population)	18	...
Telephone lines (per 100 inhabitants)	1.1[f]	1.4[d]

Social indicators	1990/95
Growth rate of population (% per annum)	3.0
Age group 0-14 years (%)	45.8
Age group 60+ years (women and men, %)	2.2/2.0
Life expectancy at birth (women and men, years)	51/47
Infant mortality rate (per 1,000 births)	112
Total fertility rate (births per woman)	7
Urban population (%)	83
Urban population growth rate (% per annum)	3.5
Rural population growth rate (% per annum)	0.7
Foreign-born (1985,%)	5.5
Refugees	28000
Government education expenditure (% of GDP)	4
Primary-secondary gross enrolment ratio (f and m, per 100)	26/37
Television receivers (per 1,000 population)	53
Parliamentary seats (women and men, %)	0/100

Environment	1990/91
Threatened species	11
Forested area (%)	0.3
CO_2 emissions (10,000 Mt)	98
Energy consumption (1,000 Mt coal equiv)	375
Precipitation (mm)	177[f]
Average temperature (January and July, centigrade)	25.0/36.7[e]

a Capital city. b 1988. c 1992. d 1991. e 1980. f 1987.

Dominica

Region	Caribbean
Location (longitude, latitude)	15°18'N 61°23'W[a]
Currency	East Caribbean dollar
Population (1995 est., in thousands)	74
Surface area (square kms)	751
Population density (pop. per square km)	111
Sex ratio (females per 100 males)	101
Largest city (pop. in thousands)	Roseau (8)
UN membership date	18-Dec-1978
Major language(s)	English, Creole French

Economic indicators	1985	1994
Exchange rate (US$)	2.70	2.70
Consumer price index (1980=100)	128	180[b]
Balance of payments, current account (million US$)	−6	−25[c]
Tourist arrivals (in thousands)	34[d]	47[c]
GDP (million US$)	99	177[e]
GDP (per capita US$)	1350	2463[e]
Long-term rate of change in GDP (% per annum)	1.3	1.8[e]
Gross fixed capital formation (% of GDP)	28.5	40.2[e]
Agricultural production index (1979-1981=100)	173[d]	153[c]
Food production index (1979-1981=100)	173[d]	153[c]
Commercial energy production (1,000 Mt coal equiv)	1[f]	2[e]
Motor vehicles (per 1,000 population)	65	112[g]
Telephone lines (per 100 inhabitants)	16.2[g]	19.4[e]

Social indicators	1990/95
Age group 0-14 years (%)	39.8[f]
Age group 60+ years (women and men, %)	5.8/4.3[h]
Infant mortality rate (per 1,000 births)	18
Contraceptive use (% of currently married women)	50[i]
Foreign-born (1985,%)	2.7
Government education expenditure (% of GDP)	5[j]
Television receivers (per 1,000 population)	72
Intentional homicides (1986, per 100,000 population)	9
Parliamentary seats (women and men, %)	13/87

Environment	1990/91
Threatened species	6
Forested area (%)	41.3
CO_2 emissions (10,000 Mt)	16
Energy consumption (1,000 Mt coal equiv)	417

a Capital city. b December 1993. c 1992. d 1988. e 1991. f 1980. g 1990. h 1981. i 1987. j 1989.

Dominican Republic

Region	Caribbean
Location (longitude, latitude)	18°47'N 69°90'W
Currency	Dominican peso
Population (1995 est., in thousands)	7915
Surface area (square kms)	48734
Population density (pop. per square km)	150
Sex ratio (females per 100 males)	97
Largest city (pop. in thousands)	Santo Domingo (2203)
UN membership date	24-Oct-1945
Major language(s)	Spanish

Economic indicators	1985	1994
Exchange rate (US$)	2.94	13.51
Consumer price index (1980=100)	212	...
Balance of payments, current account (million US$)	-108	-393[a]
Tourist arrivals (in thousands)	1116[b]	1524[a]
GDP (million US$)	4488	7202[c]
GDP (per capita US$)	700	984[c]
Long-term rate of change in GDP (% per annum)	-2.6	-0.6[c]
Gross fixed capital formation (% of GDP)	20.4	18.8[c]
Economically active female population (%)	12.1[d]	16.1
Economically active male population (%)	85.6[d]	85.0
Annual growth of econ. active female pop. (%)	5.0[e]	...
Annual growth of econ. active male pop. (%)	3.1[e]	...
Agricultural production index (1979-1981=100)	120[b]	125[a]
Food production index (1979-1981=100)	122[b]	132[a]
Commercial energy production (1,000 Mt coal equiv)	71[d]	104[c]
Motor vehicles (per 1,000 population)	24	...
Telephone lines (per 100 inhabitants)	2.9[f]	5.6[c]

Social indicators	1990/95
Growth rate of population (% per annum)	2.0
Age group 0-14 years (%)	36.3
Age group 60+ years (women and men, %)	3.0/2.9
Life expectancy at birth (women and men, years)	70/65
Infant mortality rate (per 1,000 births)	57
Total fertility rate (births per woman)	3
Contraceptive use (% of currently married women)	56
Urban population (%)	65
Urban population growth rate (% per annum)	3.3
Rural population growth rate (% per annum)	-0.3
Foreign-born (1985, %)	0.8
Refugees	500
Government education expenditure (% of GDP)	2[g]
Primary-secondary gross enrolment ratio (f and m, per 100)	94/85[h]
Third-level students (per 100,000 population)	1929[h]
Newspaper circulation (per 1,000 population)	32
Television receivers (per 1,000 population)	84
Intentional homicides (1986, per 100,000 population)	5[i]
Parliamentary seats (women and men, %)	12/88

Environment	1990/91
Threatened species	33
Forested area (%)	12.6
CO_2 emissions (10,000 Mt)	1709
Energy consumption (1,000 Mt coal equiv)	361
Precipitation (mm)	1386[f]
Average temperature (January and July, centigrade)	23.9/26.7[d]

a 1992. b 1988. c 1991. d 1980. e 1990. f 1987. g 1986. h 1985.
i Source: DYB92.

East Timor

Region	South-eastern Asia
Location (longitude, latitude)	8°33'S 125°35'E[a]
Currency	rupiah
Population (1995 est., in thousands)	835
Surface area (square kms)	14874
Population density (pop. per square km)	51
Sex ratio (females per 100 males)	97
Largest city (pop. in thousands)	Dili (99)[a]
Major language(s)	Tetum, Mambai, Bunak

Economic indicators	1985	1994
Economically active female population (%)	12.4[b]	16.4
Economically active male population (%)	89.9[b]	90.0
Annual growth of econ. active female pop. (%)	3.0[c]	...
Annual growth of econ. active male pop. (%)	1.4[c]	...

Social indicators	1990/95
Growth rate of population (% per annum)	2.0
Age group 0-14 years (%)	41.8
Age group 60+ years (women and men, %)	2.8/2.6
Life expectancy at birth (women and men, years)	46/44
Infant mortality rate (per 1,000 births)	150
Total fertility rate (births per woman)	5
Urban population (%)	15
Urban population growth rate (% per annum)	4.8
Rural population growth rate (% per annum)	1.5
Foreign-born (1985,%)	0.8

Environment	1990/91
Forested area (%)	74.0

a Capital city. b 1980. c 1990.

Ecuador

Region	South America
Location (longitude, latitude)	2°12'S 79°53'W
Currency	sucre
Population (1995 est., in thousands)	11822
Surface area (square kms)	283561
Population density (pop. per square km)	38
Sex ratio (females per 100 males)	99
Largest city (pop. in thousands)	Guayaquil (1687)
UN membership date	21-Dec-1945
Major language(s)	Spanish, Quechua

Economic indicators	1985	1994
Exchange rate (US$)	95.75	2260.50
Consumer price index (1980=100)	290	8423
Balance of payments, current account (million US$)	76	−467[a]
Tourist arrivals (in thousands)	347[b]	403[c]
GDP (million US$)	15958	11612[a]
GDP (per capita US$)	1714	1075[a]
Long-term rate of change in GDP (% per annum)	4.3	4.4[b]
Gross fixed capital formation (% of GDP)	16.1	17.5[d]
Economically active female population (%)	19.4[e]	19.5
Economically active male population (%)	83.0[e]	81.1
Annual growth of econ. active female pop. (%)	4.0[d]	...
Annual growth of econ. active male pop. (%)	2.9[d]	...
Labour force in industry (%)	29.2[f]	23.7[a]
Labour force in agriculture (%)	2.5[f]	7.9[a]
Agricultural production index (1979-1981=100)	129[b]	154[c]
Food production index (1979-1981=100)	125[b]	153[c]
Commercial energy production (1,000 Mt coal equiv)	15055[e]	23348[a]
Motor vehicles (per 1,000 population)	32	37[a]
Telephone lines (per 100 inhabitants)	3.4[f]	4.5[a]

Largest export industries		Major trading partners		1992
	(% of exports)		(% of exports)	(% of imports)
Agriculture	46	United States	47	United States 33
Mining, quarrying	42	Korea, Rep.	12	Japan 13
Food, beverages, tobacco	5	Chile	5	Germany 5

Social indicators	1990/95
Growth rate of population (% per annum)	2.3
Age group 0-14 years (%)	37.1
Age group 60+ years (women and men, %)	3.2/2.8
Life expectancy at birth (women and men, years)	69/65
Infant mortality rate (per 1,000 births)	57
Total fertility rate (births per woman)	4
Contraceptive use (% of currently married women)	53[g]
Urban population (%)	61
Urban population growth rate (% per annum)	3.8
Rural population growth rate (% per annum)	0.2
Foreign-born (1985,%)	0.8
Refugees	200
Government education expenditure (% of GDP)	2
Primary-secondary gross enrolment ratio (f and m, per 100)	89/89[f]
Third-level students (per 100,000 population)	1942
Newspaper circulation (per 1,000 population)	87
Television receivers (per 1,000 population)	84
Intentional homicides (1986, per 100,000 population)	10[h]
Parliamentary seats (women and men, %)	5/95

Environment	1990/91
Threatened species	117
Forested area (%)	38.4
CO_2 emissions (10,000 Mt)	4854
Energy consumption (1,000 Mt coal equiv)	756
Precipitation (mm)	1100[f]
Average temperature (January and July, centigrade)	25.5/23.5[e]

a 1991. b 1988. c 1992. d 1990. e 1980. f 1987. g 1989. h Source: DYB92.

Egypt

Region	Northern Africa
Location (longitude, latitude)	30°06'N 31°25'E
Currency	Egyptian pound
Population (1995 est., in thousands)	58519
Surface area (square kms)	1001449
Population density (pop. per square km)	55
Sex ratio (females per 100 males)	97
Largest city (pop. in thousands)	Cairo (8633)
UN membership date	24-Oct-1945
Major language(s)	Arabic

Economic indicators	1985	1994
Exchange rate (US$)	0.70	3.39
Consumer price index (1980=100)	477[a]	775[b]
Industrial production index (1980 = 100)	152	136[a]
Balance of payments, current account (million US$)	−2166	2812[c]
Tourist arrivals (in thousands)	1833[d]	2944[c]
GDP (million US$)	52311	32946[e]
GDP (per capita US$)	1125	614[e]
Long-term rate of change in GDP (% per annum)	8.1	5.3[e]
Economically active female population (%)	7.1[f]	9.2
Economically active male population (%)	79.8[f]	79.2
Annual growth of econ. active female pop. (%)	4.6[a]	...
Annual growth of econ. active male pop. (%)	2.4[a]	...
Labour force in industry (%)	20.1[f]	20.7[a]
Labour force in agriculture (%)	42.4[f]	39.0[a]
Agricultural production index (1979-1981=100)	127[d]	147[c]
Food production index (1979-1981=100)	137[d]	159[c]
Commercial energy production (1,000 Mt coal equiv)	45949[f]	79151[e]
Motor vehicles (per 1,000 population)	26	28[e]
Telephone lines (per 100 inhabitants)	2.3[g]	3.6[e]

Largest export industries		Major trading partners			1992
	(% of exports)		(% of exports)		(% of imports)
Mining, quarrying	39	Italy	14	United States	18
Textiles	19	Israel	10	Germany	10
Chemicals	11	United States	9	Italy	7

Social indicators	1990/95
Growth rate of population (% per annum)	2.2
Age group 0-14 years (%)	37.9
Age group 60+ years (women and men, %)	3.5/2.9
Life expectancy at birth (women and men, years)	63/60
Infant mortality rate (per 1,000 births)	57
Total fertility rate (births per woman)	4
Contraceptive use (% of currently married women)	45
Urban population (%)	45
Urban population growth rate (% per annum)	2.6
Rural population growth rate (% per annum)	1.9
Foreign-born (1985,%)	0.4
Refugees	5500
Government education expenditure (% of GDP)	3
Primary-secondary gross enrolment ratio (f and m, per 100)	81/98
Third-level students (per 100,000 population)	1698
Newspaper circulation (per 1,000 population)	57
Television receivers (per 1,000 population)	116
Intentional homicides (1986, per 100,000 population)	2
Parliamentary seats (women and men, %)	2/98

Environment	1990/91
Threatened species	37
CO_2 emissions (10,000 Mt)	22289
Energy consumption (1,000 Mt coal equiv)	713
Precipitation (mm)	22[g]
Average temperature (January and July, centigrade)	14.0/28.9[f]

a 1990. b May 1994. c 1992. d 1988. e 1991. f 1980. g 1987.

El Salvador

Region	Central America
Location (longitude, latitude)	13°70'N 89°19'W[a]
Currency	El Salvador colone
Population (1995 est., in thousands)	5768
Surface area (square kms)	21041
Population density (pop. per square km)	256
Sex ratio (females per 100 males)	104
Largest city (pop. in thousands)	San Salvador (591)[a]
UN membership date	24-Oct-1945
Major language(s)	Spanish

Economic indicators	1985	1994
Exchange rate (US$)	2.50	8.76
Consumer price index (1980=100)	198	939[b]
Industrial production index (1980 = 100)	91	104[c]
Balance of payments, current account (million US$)	−189	−213[c]
Tourist arrivals (in thousands)	134[e]	314[f]
GDP (million US$)	5732	5915[d]
GDP (per capita US$)	1210	1120[d]
Long-term rate of change in GDP (% per annum)	2.0	3.5[d]
Gross fixed capital formation (% of GDP)	12.0	13.5[d]
Economically active female population (%)	28.8[g]	28.7
Economically active male population (%)	86.8[g]	83.2
Annual growth of econ. active female pop. (%)	4.3[c]	...
Annual growth of econ. active male pop. (%)	3.0[c]	...
Labour force in industry (%)	21.0[g]	22.7[f]
Labour force in agriculture (%)	37.5[g]	35.8[f]
Agricultural production index (1979-1981=100)	79[e]	90[f]
Food production index (1979-1981=100)	107[e]	115[f]
Commercial energy production (1,000 Mt coal equiv)	580[g]	721[d]
Motor vehicles (per 1,000 population)	25	23[c]
Telephone lines (per 100 inhabitants)	2.0[h]	2.5[d]

Social indicators	1990/95
Growth rate of population (% per annum)	2.2
Age group 0-14 years (%)	40.7
Age group 60+ years (women and men, %)	3.4/2.8
Life expectancy at birth (women and men, years)	69/64
Infant mortality rate (per 1,000 births)	46
Total fertility rate (births per woman)	4
Contraceptive use (% of currently married women)	47[e]
Urban population (%)	47
Urban population growth rate (% per annum)	3.2
Rural population growth rate (% per annum)	1.4
Foreign-born (1985, %)	0.6
Refugees	19900
Government education expenditure (% of GDP)	2
Primary-secondary gross enrolment ratio (f and m, per 100)	67/66[i]
Third-level students (per 100,000 population)	1564[i]
Newspaper circulation (per 1,000 population)	88
Television receivers (per 1,000 population)	92
Intentional homicides (1986, per 100,000 population)	28[j]
Parliamentary seats (women and men, %)	8/92

Environment	1990/91
Threatened species	11
Forested area (%)	4.9
CO_2 emissions (10,000 Mt)	691
Energy consumption (1,000 Mt coal equiv)	331
Precipitation (mm)	1775[h]
Average temperature (January and July, centigrade)	22.1/23.0[g]

a Capital city. b April 1994. c 1990. d 1991. e 1988. f 1992. g 1980. h 1987. i 1989. j Source: DYB92.

Equatorial Guinea

Region	Middle Africa
Location (longitude, latitude)	3°45'N 8°47'E[a]
Currency	CFA franc
Population (1995 est., in thousands)	400
Surface area (square kms)	28051
Population density (pop. per square km)	13
Sex ratio (females per 100 males)	103
Largest city (pop. in thousands)	Malabo (31)[a]
UN membership date	12-Nov-1968
Major language(s)	Spanish, Beti (Fang variety)

Economic indicators	1985	1994
Exchange rate (US$)	378.05	528.15
Balance of payments, current account (million US$)	−19[b]	−25[c]
GDP (million US$)	85	165[c]
GDP (per capita US$)	272	457[c]
Long-term rate of change in GDP (% per annum)	6.6	3.4[c]
Gross fixed capital formation (% of GDP)	12.0	18.4[c]
Economically active female population (%)	55.5[d]	52.0
Economically active male population (%)	85.7[d]	82.1
Annual growth of econ. active female pop. (%)	1.1[b]	...
Annual growth of econ. active male pop. (%)	1.6[b]	...
Commercial energy production (1,000 Mt coal equiv)	0[d]	0[c]
Telephone lines (per 100 inhabitants)	0.4[e]	0.4[c]

Social indicators	1990/95
Growth rate of population (% per annum)	2.6
Age group 0-14 years (%)	43.3
Age group 60+ years (women and men, %)	3.8/2.8
Life expectancy at birth (women and men, years)	50/46
Infant mortality rate (per 1,000 births)	117
Total fertility rate (births per woman)	6
Urban population (%)	31
Urban population growth rate (% per annum)	3.7
Rural population growth rate (% per annum)	2.1
Foreign-born (1985, %)	1.5
Government education expenditure (% of GDP)	1[f]
Newspaper circulation (per 1,000 population)	6
Television receivers (per 1,000 population)	9

Environment	1990/91
Threatened species	25
Forested area (%)	46.2
CO_2 emissions (10,000 Mt)	33
Energy consumption (1,000 Mt coal equiv)	164

a Capital city. b 1990. c 1991. d 1980. e 1987. f 1988.

Eritrea

Region Eastern Africa
Location (longitude, latitude) 15°20'N 38°58'E[a]
Currency birr
Largest city (pop. in thousands) Asmara (358)
UN membership date 28-May-1992

Economic indicators	1985	1994
Economically active female population (%)	61.1[b]	53.3
Economically active male population (%)	90.4[b]	88.7

Environment	1990/91
Threatened species	3

a Capital city. b 1980.

Estonia

Region Northern Europe
Location (longitude, latitude) 59°25'N 24°48'E[a]
Currency rouble
Surface area (square kms) 45100
Largest city (pop. in thousands) Tallinn (498)[a]
UN membership date 17-Sep-1991
Major language(s) Estonian, Russian

Economic indicators	1985	1994
Exchange rate (US$)	...	12.60
Consumer price index (1980=100)	100[b]	157
GDP (million US$)	5825	10215[c]
GDP (per capita US$)	3792	6449[c]
Long-term rate of change in GDP (% per annum)	−8.7	−13.6[c]
Gross fixed capital formation (% of GDP)	30.4	20.1[c]
Economically active female population (%)	59.4[d]	58.6
Economically active male population (%)	79.5[d]	79.2
Labour force in industry (%)	43.0[d]	42.4[c]
Labour force in agriculture (%)	13.9[d]	12.8[c]
Motor vehicles (per 1,000 population)	114	166[c]
Telephone lines (per 100 inhabitants)	18.6[e]	21.0[c]

Social indicators	1990/95
Urban population (%)	73
Urban population growth rate (% per annum)	0.2
Rural population growth rate (% per annum)	−1.1
Television receivers (per 1,000 population)	347
Intentional homicides (1986, per 100,000 population)	11[f]

Environment	1990/91
Threatened species	27

a Capital city. b 1990. c 1991. d 1980. e 1987. f Source: DYB92.

Ethiopia

Region Eastern Africa
Location (longitude, latitude) 9°03'N 38°75'E
Currency birr
Population (1995 est., in thousands) 58039
Surface area (square kms) 1221900
Population density (pop. per square km) 44
Sex ratio (females per 100 males) 100
Largest city (pop. in thousands) Addis Ababa (1808)
UN membership date 13-Nov-1945
Major language(s) Amharic, Oromo (Galla), Tigrinya

Economic indicators	1985	1994
Exchange rate (US$)	2.07	5.60
Consumer price index (1980=100)	144	256[a]
Industrial production index (1980 = 100)	143	...
Balance of payments, current account (million US$)	106	−120[b]
Tourist arrivals (in thousands)	76[c]	83[b]
GDP (million US$)	4778	6592[d]
GDP (per capita US$)	118	137[d]
Long-term rate of change in GDP (% per annum)	−7.0	−4.1[d]
Gross fixed capital formation (% of GDP)	14.0	10.4[d]
Economically active female population (%)	56.7[e]	49.5
Economically active male population (%)	89.6[e]	88.3
Annual growth of econ. active female pop. (%)	1.7[f]	...
Annual growth of econ. active male pop. (%)	2.3[f]	...
Agricultural production index (1979-1981=100)	105[c]	113[b]
Food production index (1979-1981=100)	106[c]	115[b]
Commercial energy production (1,000 Mt coal equiv)	78[e]	179[d]
Motor vehicles (per 1,000 population)	1	1[d]
Telephone lines (per 100 inhabitants)	0.2[g]	0.3[d]

Social indicators	1990/95
Growth rate of population (% per annum)	3.0
Age group 0-14 years (%)	46.5
Age group 60+ years (women and men, %)	2.5/2.0
Life expectancy at birth (women and men, years)	49/45
Infant mortality rate (per 1,000 births)	122
Total fertility rate (births per woman)	7
Contraceptive use (% of currently married women)	4
Urban population (%)	13
Urban population growth rate (% per annum)	4.8
Rural population growth rate (% per annum)	2.8
Foreign-born (1985,%)	0.3
Refugees	431800
Government education expenditure (% of GDP)	4
Primary-secondary gross enrolment ratio (f and m, per 100)	17/22
Third-level students (per 100,000 population)	70[h]
Newspaper circulation (per 1,000 population)	1[c]
Television receivers (per 1,000 population)	3

Environment	1990/91
Threatened species	52
Forested area (%)	22.2
CO_2 emissions (10,000 Mt)	771
Energy consumption (1,000 Mt coal equiv)	27
Precipitation (mm)	1089[g]
Average temperature (January and July, centigrade)	16.1/15.1[e]

a April 1994. b 1992. c 1988. d 1991. e 1980. f 1990. g 1987. h 1989.

Fiji

Region	Oceania-Melanesia
Location (longitude, latitude)	18°14'S 178°43'E[a]
Currency	Fiji dollar
Population (1995 est., in thousands)	762
Surface area (square kms)	18274
Population density (pop. per square km)	42
Sex ratio (females per 100 males)	97
Largest city (pop. in thousands)	Suva (200)[a]
UN membership date	13-Oct-1970
Major language(s)	English, Fijian, Hindi

Economic indicators	1985	1994
Exchange rate (US$)	1.12	1.44
Consumer price index (1980=100)	140	164
Industrial production index (1980 = 100)	103	141[b]
Balance of payments, current account (million US$)	–12	–13[c]
Tourist arrivals (in thousands)	208[d]	279[c]
GDP (million US$)	1141	1311[e]
GDP (per capita US$)	1632	1791[e]
Long-term rate of change in GDP (% per annum)	–3.9	–0.2[e]
Gross fixed capital formation (% of GDP)	18.2	...
Economically active female population (%)	18.0[f]	22.8
Economically active male population (%)	85.5[f]	82.9
Annual growth of econ. active female pop. (%)	5.7[g]	...
Annual growth of econ. active male pop. (%)	2.2[g]	...
Labour force in industry (%)	34.5[f]	36.5[c]
Labour force in agriculture (%)	3.2[f]	2.5[c]
Agricultural production index (1979-1981=100)	98[d]	110[c]
Food production index (1979-1981=100)	98[d]	110[c]
Commercial energy production (1,000 Mt coal equiv)	47[g]	47[e]
Motor vehicles (per 1,000 population)	85	105[e]
Telephone lines (per 100 inhabitants)	4.8[h]	6.2[e]

Largest export industries	Major trading partners		1992
(% of exports)	(% of exports)	(% of imports)	
Food, beverages, tobacco 48	United Kingdom 33	Australia	32
Textiles 22	United States 15	New Zealand	17
Chemicals 11	Australia 9	Japan	10

Social indicators	1990/95
Growth rate of population (% per annum)	1.0
Age group 0-14 years (%)	35.0
Age group 60+ years (women and men, %)	3.3/2.9
Life expectancy at birth (women and men, years)	74/70
Infant mortality rate (per 1,000 births)	23
Total fertility rate (births per woman)	3
Contraceptive use (% of currently married women)	40[i,j]
Urban population (%)	41
Urban population growth rate (% per annum)	1.7
Rural population growth rate (% per annum)	0.5
Foreign-born (1985, %)	1.9
Government education expenditure (% of GDP)	5[k]
Primary-secondary gross enrolment ratio (f and m, per 100)	90/90[i]
Third-level students (per 100,000 population)	406[d]
Newspaper circulation (per 1,000 population)	37
Television receivers (per 1,000 population)	15
Intentional homicides (1986, per 100,000 population)	3
Parliamentary seats (women and men, %)	1/99

Environment	1990/91
Threatened species	22
Forested area (%)	64.9
CO_2 emissions (10,000 Mt)	188
Energy consumption (1,000 Mt coal equiv)	478

a Capital city. b June 1994. c 1992. d 1988. e 1991. f 1980. g 1990. h 1987. i 1986. j Source: NATHR. k 1989.

Finland

Region	Northern Europe
Location (longitude, latitude)	60°16'N 24°94'E
Currency	markkaa
Population (1995 est., in thousands)	5046
Surface area (square kms)	3381445
Population density (pop. per square km)	15
Sex ratio (females per 100 males)	106
Largest city (pop. in thousands)	Helsinki (1009)
UN membership date	14-Dec-1955
Major language(s)	Finnish, Swedish

Economic indicators	1985	1994
Exchange rate (US$)	5.42	4.87
Consumer price index (1980=100)	151	214
Industrial production index (1980 = 100)	116	106[a]
Unemployment (%)	...	17.1[b]
Balance of payments, current account (million US$)	−811	−980[c]
Tourist arrivals (in thousands)	877[d]	790[e]
GDP (million US$)	54048	124541[f]
GDP (per capita US$)	11026	24933[f]
Long-term rate of change in GDP (% per annum)	3.3	−6.4[f]
Gross fixed capital formation (% of GDP)	23.9	22.1[f]
Economically active female population (%)	55.4[g]	56.8
Economically active male population (%)	70.3[g]	69.2
Annual growth of econ. active female pop. (%)	1.1[h]	...
Annual growth of econ. active male pop. (%)	0.4[h]	...
Labour force in industry (%)	34.1[g]	27.4[e]
Labour force in agriculture (%)	13.3[g]	8.5[e]
Agricultural production index (1979-1981=100)	104[d]	101[e]
Food production index (1979-1981=100)	104[d]	101[e]
Commercial energy production (1,000 Mt coal equiv)	4072[g]	10804[f]
Motor vehicles (per 1,000 population)	354	441[f]
Telephone lines (per 100 inhabitants)	47.9[i]	54.4[f]

Largest export industries		Major trading partners		1992
	(% of exports)	(% of exports)		(% of imports)
Metal manufacture	34	Germany	15	Germany 17
Paper, paper products	30	Sweden	13	Sweden 11
Chemicals	11	United Kingdom	10	United Kingdom 9

Social indicators	1990/95
Growth rate of population (% per annum)	0.3
Age group 0-14 years (%)	18.9
Age group 60+ years (women and men, %)	11.4/7.4
Life expectancy at birth (women and men, years)	80/72
Infant mortality rate (per 1,000 births)	6
Total fertility rate (births per woman)	2
Urban population (%)	60
Urban population growth rate (% per annum)	0.5
Rural population growth rate (% per annum)	−0.3
Foreign-born (1985,%)	1.0
Refugees	12000
Government education expenditure (% of GDP)	7
Primary-secondary gross enrolment ratio (f and m, per 100)	111/102
Third-level students (per 100,000 population)	3331
Newspaper circulation (per 1,000 population)	558
Television receivers (per 1,000 population)	501
Intentional homicides (1986, per 100,000 population)	3
Parliamentary seats (women and men, %)	39/61

Environment	1990/91
Threatened species	36
Forested area (%)	68.7
CO_2 emissions (10,000 Mt)	14205
Energy consumption (1,000 Mt coal equiv)	6845
Precipitation (mm)	641[i]
Average temperature (January and July, centigrade)	−6.8/17.1[g]

a July 1994. b Labour force sample surveys. c 1993. d 1988. e 1992.
f 1991. g 1980. h 1990. i 1987.

France

Region Western Europe
Location (longitude, latitude) 48°85'N 2°32'E
Currency French franc
Population (1995 est., in thousands) 57769
Surface area (square kms) 551500
Population density (pop. per square km) 103
Sex ratio (females per 100 males) 105
Largest city (pop. in thousands) Paris (9334)
UN membership date 24-Oct-1945
Major language(s) French

Economic indicators	1985	1994
Exchange rate (US$)	7.56	5.28
Consumer price index (1980=100)	158	202
Industrial production index (1980 = 100)	101	85[a]
Balance of payments, current account (million US$)	−35	4109[b]
Tourist arrivals (in thousands)	42721[c]	59590[b]
GDP (million US$)	523098	1199291[d]
GDP (per capita US$)	9482	21053[d]
Long-term rate of change in GDP (% per annum)	1.9	1.2[d]
Gross fixed capital formation (% of GDP)	19.3	20.8[d]
Economically active female population (%)	43.0[e]	44.4
Economically active male population (%)	70.8[e]	69.8
Annual growth of econ. active female pop. (%)	1.3[f]	...
Annual growth of econ. active male pop. (%)	0.5[f]	...
Labour force in industry (%)	35.5[e]	28.5[b]
Labour force in agriculture (%)	8.4[e]	5.1[b]
Agricultural production index (1979-1981=100)	106[c]	112[b]
Food production index (1979-1981=100)	106[c]	112[b]
Commercial energy production (1,000 Mt coal equiv)[g]	64570[c]	148922[d]
Motor vehicles (per 1,000 population)	440	489[d]
Telephone lines (per 100 inhabitants)	44.5[h]	51.1[d]

Largest export industries	Major trading partners			1992
	(% of exports)[g]	(% of exports)[g]		(% of imports)[g]
Metal manufacture 46	Germany	18	Germany	19
Chemicals 18	Italy	11	Italy	11
Food, beverages, tobacco 11	Belgium[i]	9	Belgium[i]	9

Social indicators	1990/95
Growth rate of population (% per annum)	0.4
Age group 0-14 years (%)	19.8
Age group 60+ years (women and men, %)	11.5/8.3
Life expectancy at birth (women and men, years)	81/73
Infant mortality rate (per 1,000 births)	7
Total fertility rate (births per woman)	2
Contraceptive use (% of currently married women)	81[c]
Urban population (%)	73
Urban population growth rate (% per annum)	0.4
Rural population growth rate (% per annum)	0.3
Foreign-born (1985,%)	10.8
Refugees	182600
Government education expenditure (% of GDP)	5
Primary-secondary gross enrolment ratio (f and m, per 100)	105/101[j]
Third-level students (per 100,000 population)	3026
Newspaper circulation (per 1,000 population)	208
Television receivers (per 1,000 population)	407
Intentional homicides (1986, per 100,000 population)	4
Parliamentary seats (women and men, %)	6/94

Environment	1990/91
Threatened species	118
Forested area (%)	26.9
CO_2 emissions (10,000 Mt)	102105[k]
Energy consumption (1,000 Mt coal equiv)[g]	5457
Precipitation (mm)	585[h]
Average temperature (January and July, centigrade)	3.1/19.0[e]

a August 1994. b 1992. c 1988. d 1991. e 1980. f 1990. g Includes Monaco. h 1987. i Includes Luxemburg. j 1989. k Includes San Marino.

French Guiana

Region	South America
Location (longitude, latitude)	4°50'N 52°22'W[a]
Currency	French franc
Population (1995 est., in thousands)	73
Surface area (square kms)	90000
Population density (pop. per square km)	1
Sex ratio (females per 100 males)	90
Largest city (pop. in thousands)	Cayenne (41)[a]
Major language(s)	French

Economic indicators	1985	1994
Consumer price index (1980=100)	170	219[b]
Telephone lines (per 100 inhabitants)	24.6[e]	32.7[d]

Largest export industries	Major trading partners		1992
(% of exports)	(% of exports)	(% of imports)	
Agriculture 53	Guadeloupe 10	Trinidad Tbg. 8	
Metal manufacture 40	Martinique 5	Italy 3	
Food, beverages, tobacco 4	United Kingdom 5	United States 3	

Social indicators	1990/95
Age group 0-14 years (%)	32.6[c]
Age group 60+ years (women and men, %)	3.6/3.3[f]
Infant mortality rate (per 1,000 births)	22
Urban population (%)	77
Urban population growth rate (% per annum)	3.6
Rural population growth rate (% per annum)	1.5
Foreign-born (1985,%)	50.3
Refugees	1700
Newspaper circulation (per 1,000 population)	10
Television receivers (per 1,000 population)	213

Environment	1990/91
Threatened species	20
Forested area (%)	81.1
CO_2 emissions (10,000 Mt)	190
Energy consumption (1,000 Mt coal equiv)	3257
Precipitation (mm)	3744[e]
Average temperature (January and July, centigrade)	25.1/25.1[c]

a Capital city. b July 1994. c 1980. d 1991. e 1987. f 1982.

French Polynesia

Region	Oceania-Polynesia
Location (longitude, latitude)	17°54'S 149°57'W[a]
Currency	CFP franc
Population (1995 est., in thousands)	222
Surface area (square kms)	4000
Population density (pop. per square km)	53
Sex ratio (females per 100 males)	93
Largest city (pop. in thousands)	Papeete (112)[a]
Major language(s)	French, Tahitian

Economic indicators	1985	1994
Consumer price index (1980=100)	184	212[b]
Tourist arrivals (in thousands)	135[c]	124[d]
GDP (million US$)	1388	3036[e]
GDP (per capita US$)	7978	14955[e]
Long-term rate of change in GDP (% per annum)	5.4	3.9[e]
Gross fixed capital formation (% of GDP)	35.3	21.2[f]
Economically active female population (%)	32.6[g]	37.0
Economically active male population (%)	84.5[g]	83.8
Labour force in industry (%)	20.0[h]	19.2[i]
Labour force in agriculture (%)	1.8[h]	2.2[i]
Agricultural production index (1979-1981=100)	95[c]	95[d]
Food production index (1979-1981=100)	97[c]	97[d]
Commercial energy production (1,000 Mt coal equiv)	9[f]	9[e]
Telephone lines (per 100 inhabitants)	16.0[j]	20.3[e]

Social indicators	1990/95
Growth rate of population (% per annum)	2.3
Age group 0-14 years (%)	34.7
Age group 60+ years (women and men, %)	3.2/3.2
Life expectancy at birth (women and men, years)	73/68
Infant mortality rate (per 1,000 births)	16
Total fertility rate (births per woman)	3
Urban population (%)	67
Urban population growth rate (% per annum)	3.0
Rural population growth rate (% per annum)	0.9
Foreign-born (1985,%)	14.8
Newspaper circulation (per 1,000 population)	106
Television receivers (per 1,000 population)	169

Environment	1990/91
Threatened species	35
Forested area (%)	28.8
CO_2 emissions (10,000 Mt)	168
Energy consumption (1,000 Mt coal equiv)	1493
Precipitation (mm)	1800[j]
Average temperature (January and July, centigrade)	26.0/24.1[g]

a Capital city. b August 1994. c 1988. d 1992. e 1991. f 1990. g 1980. h 1982. i 1989. j 1987.

Gabon

Region	Middle Africa
Location (longitude, latitude)	0°27'N 9°25'E[a]
Currency	CFA franc
Population (1995 est., in thousands)	1367
Surface area (square kms)	267667
Population density (pop. per square km)	5
Sex ratio (females per 100 males)	103
Largest city (pop. in thousands)	Libreville (289)[a]
UN membership date	20-Sep-1960
Major language(s)	French, Beti (Fang variety), Kota, Mbere, Mpongwe, Njebi, Punu, Teke

Economic indicators	1985	1994
Exchange rate (US$)	378.05	528.15
Consumer price index (1980=100)	159	196[b]
Balance of payments, current account (million US$)	−163	−135[c]
Tourist arrivals (in thousands)	20[d]	130[c]
GDP (million US$)	3663	4438[e]
GDP (per capita US$)	3719	3708[e]
Long-term rate of change in GDP (% per annum)	1.0	1.9[e]
Economically active female population (%)	51.7[f]	45.0
Economically active male population (%)	83.9[f]	81.6
Annual growth of econ. active female pop. (%)	0.5[g]	...
Annual growth of econ. active male pop. (%)	1.0[g]	...
Agricultural production index (1979-1981=100)	116[d]	123[e]
Food production index (1979-1981=100)	116[d]	123[e]
Commercial energy production (1,000 Mt coal equiv)	12757[f]	25009[e]
Motor vehicles (per 1,000 population)	27	...
Telephone lines (per 100 inhabitants)	1.3[h]	1.8[e]

Social indicators	1990/95
Growth rate of population (% per annum)	3.3
Age group 0-14 years (%)	35.9
Age group 60+ years (women and men, %)	5.0/4.0
Life expectancy at birth (women and men, years)	55/52
Infant mortality rate (per 1,000 births)	94
Total fertility rate (births per woman)	5
Urban population (%)	50
Urban population growth rate (% per annum)	5.1
Rural population growth rate (% per annum)	1.6
Foreign-born (1985,%)	3.1
Refugees	300
Government education expenditure (% of GDP)	5[h]
Third-level students (per 100,000 population)	377[d]
Newspaper circulation (per 1,000 population)	17
Television receivers (per 1,000 population)	37
Parliamentary seats (women and men, %)	6/94

Environment	1990/91
Threatened species	32
Forested area (%)	74.7
CO_2 emissions (10,000 Mt)	1634
Energy consumption (1,000 Mt coal equiv)	716
Precipitation (mm)	3120[h]
Average temperature (January and July, centigrade)	26.7/24.1[f]

a Capital city. b February 1994. c 1992. d 1988. e 1991. f 1980. g 1990. h 1987.

Gambia

Region	Western Africa
Location (longitude, latitude)	13°28'N 16°39'W
Currency	dalasi
Population (1995 est., in thousands)	980
Surface area (square kms)	11295
Population density (pop. per square km)	78
Sex ratio (females per 100 males)	102
Largest city (pop. in thousands)	Banjul (195)[a]
UN membership date	21-Sep-1965
Major language(s)	English, Malinke, Ful, Wolof

Economic indicators	1985	1994
Exchange rate (US$)	3.46	9.70
Consumer price index (1980=100)	188	627[b]
Balance of payments, current account (million US$)	7	37[c]
Tourist arrivals (in thousands)	102[d]	95[c]
GDP (million US$)	201	306[e]
GDP (per capita US$)	270	346[e]
Long-term rate of change in GDP (% per annum)	1.6	2.3[e]
Gross fixed capital formation (% of GDP)	7.1	5.4[f]
Economically active female population (%)	63.9[g]	56.4
Economically active male population (%)	92.2[g]	90.9
Annual growth of econ. active female pop. (%)	1.4[f]	...
Annual growth of econ. active male pop. (%)	1.9[f]	...
Labour force in industry (%)	24.9[h]	23.8[i]
Labour force in agriculture (%)	8.5[h]	7.7[i]
Agricultural production index (1979-1981=100)	122[d]	111[c]
Food production index (1979-1981=100)	122[d]	104[c]
Motor vehicles (per 1,000 population)	8	...
Telephone lines (per 100 inhabitants)	0.5[i]	1.6[e]

Social indicators	1990/95
Growth rate of population (% per annum)	2.6
Age group 0-14 years (%)	43.9
Age group 60+ years (women and men, %)	2.7/2.3
Life expectancy at birth (women and men, years)	47/43
Infant mortality rate (per 1,000 births)	132
Total fertility rate (births per woman)	6
Urban population (%)	26
Urban population growth rate (% per annum)	5.0
Rural population growth rate (% per annum)	1.8
Foreign-born (1985, %)	12.2
Refugees	3600
Government education expenditure (% of GDP)	3
Primary-secondary gross enrolment ratio (f and m, per 100)	35/54
Newspaper circulation (per 1,000 population)	2
Parliamentary seats (women and men, %)	8/92

Environment	1990/91
Threatened species	7
Forested area (%)	13.8
CO_2 emissions (10,000 Mt)	54
Energy consumption (1,000 Mt coal equiv)	106

a Capital city. b March 1994. c 1992. d 1988. e 1991. f 1990. g 1980. h 1983. i 1987.

Georgia

Region Western Asia
Location (longitude, latitude) 41°41'N 44°57'E
Surface area (square kms) 69700
Largest city (pop. in thousands) Tbilisi (1277)
UN membership date 31-July-1992
Major language(s) Georgian, Russian, Merrelian, Armenian, Azerbaijani

Economic indicators	1985	1994
GDP (million US$)	19156	12314[a]
GDP (per capita US$)	3623	2252[a]
Long-term rate of change in GDP (% per annum)	12.7	−20.3[a]
Economically active female population (%)	55.5[b]	54.6
Economically active male population (%)	76.2[b]	76.6
Telephone lines (per 100 inhabitants)	8.5[c]	10.3[a]

Social indicators	1990/95
Contraceptive use (% of currently married women)	17

Environment	1990/91
Threatened species	39

a 1991. b 1980. c 1987.

Germany

Region Western Europe
Location (longitude, latitude) 51°25'N 6°57'E
Currency deutsche mark
Population (1995 est., in thousands) 81264
Sex ratio (females per 100 males) 106
Largest city (pop. in thousands) Essen (6353)
UN membership date 18-Sep-1973
Major language(s) German

Economic indicators	1985	1994
Exchange rate (US$)	2.46a	1.55a
Consumer price index (1980=100)	100a	139
Industrial production index (1980 = 100)	104a	124b
Unemployment (%)	...	8.8a
Balance of payments, current account (million US$)	17a	−21c
Tourist arrivals (in thousands)a	14501d	15147b
GDP (million US$)	715754	1686557e
GDP (per capita US$)	15944	22436e
Long-term rate of change in GDP (% per annum)	2.3	0.4e
Gross fixed capital formation (% of GDP)a	19.5	21.0f
Economically active female population (%)	44.2g	45.1
Economically active male population (%)	75.2g	76.3
Annual growth of econ. active female pop. (%)a	0.6f	...
Annual growth of econ. active male pop. (%)a	0.4f	...
Labour force in industry (%)	44.3ag	40.0ae
Labour force in agriculture (%)	5.4ag	3.5ae
Agricultural production index (1979-1981=100)a	117d	119f
Food production index (1979-1981=100)a	117d	119f
Commercial energy production (1,000 Mt coal equiv)a	174822g	..e
Motor vehicles (per 1,000 population)a	451	527e
Telephone lines (per 100 inhabitants)a	..h	42.0e

Largest export industries		Major trading partners		1992	
	(% of exports)	(% of exports)		(% of imports)	
Metal manufacture	59	Francei	13	Francei	12
Chemicals	16	Italy	9	Netherlands	10
Textiles	6	Netherlands	8	Italy	9

Social indicators	1990/95
Growth rate of population (% per annum)	0.4
Age group 0-14 years (%)	17.1
Age group 60+ years (women and men, %)	12.3/7.8
Life expectancy at birth (women and men, years)	79/73
Infant mortality rate (per 1,000 births)	7
Total fertility rate (births per woman)	2
Contraceptive use (% of currently married women)	75
Urban population (%)	86
Urban population growth rate (% per annum)	0.7
Rural population growth rate (% per annum)	−1.2
Foreign-born (1985,%)a	7.3
Refugees	827100
Government education expenditure (% of GDP)a	4
Primary-secondary gross enrolment ratio (f and m, per 100)a	103/106j
Third-level students (per 100,000 population)a	2810j
Newspaper circulation (per 1,000 population)a	338j
Television receivers (per 1,000 population)a	483k
Intentional homicides (1986, per 100,000 population)	1l
Parliamentary seats (women and men, %)	21/79

Environment	1990/91
Threatened species	75
Forested area (%)a	29.8
CO_2 emissions (10,000 Mt)	264637
Energy consumption (1,000 Mt coal equiv)a	6375
Precipitation (mm)	720am
Average temperature (January and July, centigrade)a	0.0/17.3g

a Data refer to Germany, Federal Republic of. b 1992. c 1993. d 1988.
e 1991. f 1990. g 1980. h 1987. i Includes Monaco. j 1989. k 1985.
l Source: DYB92. m Hamburg.

Ghana

Region Western Africa
Location (longitude, latitude) 5°53'N 0°23'W
Currency cedi
Population (1995 est., in thousands) 17453
Surface area (square kms) 238533
Population density (pop. per square km) 65
Sex ratio (females per 100 males) 101
Largest city (pop. in thousands) Accra (1405)
UN membership date 08-Mar-1957
Major language(s) English, Akan, Ewe

Economic indicators	1985	1994
Exchange rate (US$)	59.99	943.40
Consumer price index (1980=100)	909	7265[a]
Industrial production index (1980 = 100)	74	111[b]
Balance of payments, current account (million US$)	-134	-378[c]
Tourist arrivals (in thousands)	114[d]	213[c]
GDP (million US$)	6346	7100[e]
GDP (per capita US$)	494	459[e]
Long-term rate of change in GDP (% per annum)	5.1	3.8[e]
Gross fixed capital formation (% of GDP)	9.5	12.2[b]
Economically active female population (%)	55.5[f]	50.6
Economically active male population (%)	82.0[f]	80.6
Annual growth of econ. active female pop. (%)	2.3[b]	...
Annual growth of econ. active male pop. (%)	2.8[b]	...
Labour force in industry (%)	26.1[f]	27.3[g]
Labour force in agriculture (%)	16.3[f]	8.5[g]
Agricultural production index (1979-1981=100)	148[d]	161[c]
Food production index (1979-1981=100)	150[d]	163[c]
Commercial energy production (1,000 Mt coal equiv)	774[f]	750[e]
Motor vehicles (per 1,000 population)	8	8[b]
Telephone lines (per 100 inhabitants)	0.3[h]	0.3[e]

Social indicators	1990/95
Growth rate of population (% per annum)	3.0
Age group 0-14 years (%)	45.3
Age group 60+ years (women and men, %)	2.5/2.1
Life expectancy at birth (women and men, years)	58/54
Infant mortality rate (per 1,000 births)	81
Total fertility rate (births per woman)	6
Contraceptive use (% of currently married women)	13[d]
Urban population (%)	36
Urban population growth rate (% per annum)	4.3
Rural population growth rate (% per annum)	2.3
Foreign-born (1985,%)	1.5
Refugees	12100
Government education expenditure (% of GDP)	3
Primary-secondary gross enrolment ratio (f and m, per 100)	50/66[g]
Third-level students (per 100,000 population)	127[g]
Newspaper circulation (per 1,000 population)	13
Television receivers (per 1,000 population)	15
Parliamentary seats (women and men, %)	8/93

Environment	1990/91
Threatened species	31
Forested area (%)	33.8
CO_2 emissions (10,000 Mt)	943
Energy consumption (1,000 Mt coal equiv)	124
Precipitation (mm)	787[h]
Average temperature (January and July, centigrade)	27.3/24.6[f]

a July 1994. b 1990. c 1992. d 1988. e 1991. f 1980. g 1989. h 1987.

Greece

Region	Southern Europe
Location (longitude, latitude)	37°97'N 23°74'E
Currency	drachma
Population (1995 est., in thousands)	10253
Surface area (square kms)	131990
Population density (pop. per square km)	76
Sex ratio (females per 100 males)	103
Largest city (pop. in thousands)	Athens (3461)
UN membership date	25-Oct-1945
Major language(s)	Greek

Economic indicators	1985	1994
Exchange rate (US$)	147.76	236.00
Consumer price index (1980=100)	256	1028
Industrial production index (1980 = 100)	107	102[a]
Balance of payments, current account (million US$)	−3276	−747[b]
Tourist arrivals (in thousands)	7923[c]	9331[d]
GDP (million US$)	33433	70574[e]
GDP (per capita US$)	3366	6951[e]
Long-term rate of change in GDP (% per annum)	3.1	1.8[e]
Gross fixed capital formation (% of GDP)	19.1	18.2[e]
Economically active female population (%)	24.6[f]	24.9
Economically active male population (%)	74.7[f]	71.6
Annual growth of econ. active female pop. (%)	0.9[g]	...
Annual growth of econ. active male pop. (%)	0.6[g]	...
Labour force in industry (%)	29.0[h]	27.5[e]
Labour force in agriculture (%)	30.7[h]	22.2[e]
Agricultural production index (1979-1981=100)	113[c]	120[d]
Food production index (1979-1981=100)	109[c]	115[d]
Commercial energy production (1,000 Mt coal equiv)	4760[f]	11661[e]
Motor vehicles (per 1,000 population)	190	256[e]
Telephone lines (per 100 inhabitants)	34.6[i]	41.3[e]

Social indicators	1990/95
Growth rate of population (% per annum)	0.3
Age group 0-14 years (%)	17.3
Age group 60+ years (women and men, %)	12.1/9.7
Life expectancy at birth (women and men, years)	80/75
Infant mortality rate (per 1,000 births)	8
Total fertility rate (births per woman)	1
Urban population (%)	65
Urban population growth rate (% per annum)	1.1
Rural population growth rate (% per annum)	−1.2
Foreign-born (1985, %)	2.3
Refugees	8500
Government education expenditure (% of GDP)	3[j]
Primary-secondary gross enrolment ratio (f and m, per 100)	99/100[c]
Third-level students (per 100,000 population)	2200[c]
Newspaper circulation (per 1,000 population)	139[c]
Television receivers (per 1,000 population)	197
Intentional homicides (1986, per 100,000 population)	2
Parliamentary seats (women and men, %)	5/95

Environment	1990/91
Threatened species	60
Forested area (%)	19.8
CO_2 emissions (10,000 Mt)	19887
Energy consumption (1,000 Mt coal equiv)	3169
Precipitation (mm)	402[i]
Average temperature (January and July, centigrade)	9.3/27.6[f]

a August 1994. b 1993. c 1988. d 1992. e 1991. f 1980. g 1990. h 1981. i 1987. j 1989.

Grenada

Region Caribbean
Location (longitude, latitude) 12°03'N 61°45'W
Currency East Caribbean dollar
Population (1995 est., in thousands) 89
Surface area (square kms) 344
Population density (pop. per square km) 244
Sex ratio (females per 100 males) 107
Largest city (pop. in thousands) St. George's (5)[a]
UN membership date 17-Sep-1974
Major language(s) English

Economic indicators	1985	1994
Exchange rate (US$)	2.70	2.70
Consumer price index (1980=100)	147	127[b]
Balance of payments, current account (million US$)	2	−25[c]
Tourist arrivals (in thousands)	59[d]	88[c]
GDP (million US$)	115	210[e]
GDP (per capita US$)	1281	2309[e]
Long-term rate of change in GDP (% per annum)	8.2	3.0[e]
Gross fixed capital formation (% of GDP)	31.2	40.0[e]
Agricultural production index (1979-1981=100)	82[d]	80[c]
Food production index (1979-1981=100)	82[d]	80[c]
Telephone lines (per 100 inhabitants)	7.8[g]	17.8[e]

Social indicators	1990/95
Age group 0-14 years (%)	38.6[f]
Age group 60+ years (women and men, %)	6.0/4.0[h]
Infant mortality rate (per 1,000 births)	15[i]
Contraceptive use (% of currently married women)	54
Foreign-born (1985,%)	3.3
Government education expenditure (% of GDP)	5[j]
Television receivers (per 1,000 population)	330

Environment	1990/91
Threatened species	5
Forested area (%)	8.8
CO_2 emissions (10,000 Mt)	33
Energy consumption (1,000 Mt coal equiv)	637

a Capital city. b June 1994. c 1992. d 1988. e 1991. f 1980. g 1987.
h 1981. i 1975-80 data. j 1986.

Guadeloupe

Region	Caribbean
Location (longitude, latitude)	16°16'N 61°31'W[a]
Currency	French franc
Population (1995 est., in thousands)	414
Surface area (square kms)	1705
Population density (pop. per square km)	202
Sex ratio (females per 100 males)	105
Largest city (pop. in thousands)	Pointe-a-Pitre (26)[a]
Major language(s)	French

Economic indicators	1985	1994
Consumer price index (1980=100)	161	202[b]
Tourist arrivals (in thousands)	329[c]	300[d]
GDP (million US$)	1091	2610[e]
GDP (per capita US$)	3073	6693[e]
Long-term rate of change in GDP (% per annum)	−0.8	−4.0[e]
Gross fixed capital formation (% of GDP)	21.8	...
Economically active female population (%)	49.0[f]	55.5
Economically active male population (%)	72.2[f]	74.7
Annual growth of econ. active female pop. (%)	2.4[g]	...
Annual growth of econ. active male pop. (%)	1.4[g]	...
Agricultural production index (1979-1981=100)	93[c]	75[d]
Food production index (1979-1981=100)	93[c]	75[d]
Motor vehicles (per 1,000 population)	332	318[e]
Telephone lines (per 100 inhabitants)	26.4[h]	32.1[e]

Largest export industries		Major trading partners			1992
	(% of exports)		(% of exports)		(% of imports)
Agriculture	43	Martinique	16	Martinique	4
Food, beverages, tobacco	34	Portugal	7	United States	4
Metal manufacture	17	Sweden	4	Italy	4

Social indicators	1990/95
Growth rate of population (% per annum)	1.2
Age group 0-14 years (%)	25.6
Age group 60+ years (women and men, %)	6.5/5.1
Life expectancy at birth (women and men, years)	78/71
Infant mortality rate (per 1,000 births)	12
Total fertility rate (births per woman)	2
Urban population (%)	51
Urban population growth rate (% per annum)	2.4
Rural population growth rate (% per annum)	1.0
Foreign-born (1985,%)	12.7
Newspaper circulation (per 1,000 population)	50
Television receivers (per 1,000 population)	263

Environment	1990/91
Threatened species	8
Forested area (%)	38.6
CO_2 emissions (10,000 Mt)	309
Energy consumption (1,000 Mt coal equiv)	1248
Precipitation (mm)	1814[h]
Average temperature (January and July, centigrade)	23.4/26.7[f]

a Capital city. b August 1994. c 1988. d 1992. e 1991. f 1980. g 1990. h 1987.

Guam

Region	Oceania-Micronesia
Location (longitude, latitude)	13°28'N 144°45'E[a]
Currency	US dollar
Population (1995 est., in thousands)	146
Surface area (square kms)	541
Population density (pop. per square km)	220
Sex ratio (females per 100 males)	92
Largest city (pop. in thousands)	Agana (3)[a]
Major language(s)	English, Chamorro, Tagalog

Economic indicators	1985	1994
Consumer price index (1980=100)	148	339
Tourist arrivals (in thousands)	586[b]	877[c]
Economically active female population (%)	33.5[d]	36.5
Economically active male population (%)	84.4[d]	84.0
Annual growth of econ. active female pop. (%)	4.2[e]	...
Annual growth of econ. active male pop. (%)	−0.8[e]	...
Labour force in industry (%)	9.4[f]	13.8[g]
Labour force in agriculture (%)	0.3[f]	0.5[g]
Motor vehicles (per 1,000 population)	642	...
Telephone lines (per 100 inhabitants)	29.1[e]	31.8[h]

Social indicators	1990/95
Growth rate of population (% per annum)	1.7
Age group 0-14 years (%)	30.1
Age group 60+ years (women and men, %)	4.1/4.1
Life expectancy at birth (women and men, years)	79/73
Infant mortality rate (per 1,000 births)	8
Total fertility rate (births per woman)	3
Urban population (%)	58
Urban population growth rate (% per annum)	3.6
Rural population growth rate (% per annum)	−0.7
Foreign-born (1985,%)	49.5
Newspaper circulation (per 1,000 population)	161
Television receivers (per 1,000 population)	658

Environment	1990/91
Threatened species	66
Forested area (%)	18.2
CO_2 emissions (10,000 Mt)	409
Energy consumption (1,000 Mt coal equiv)	5184
Precipitation (mm)	2249[g]
Average temperature (January and July, centigrade)	25.6/26.4[d]

a Capital city. b 1988. c 1992. d 1980. e 1990. f 1982. g 1987. h 1991.

Guatemala

Region Central America
Location (longitude, latitude) 14°35'N 90°32'W
Currency quetzale
Population (1995 est., in thousands) 10621
Surface area (square kms) 108889
Population density (pop. per square km) 87
Sex ratio (females per 100 males) 98
Largest city (pop. in thousands) Guatemala City (842)
UN membership date 21-Nov-1945
Major language(s) Spanish, Quich

Economic indicators	1985	1994
Exchange rate (US$)	1.00	5.81
Consumer price index (1980=100)	119	542[a]
Balance of payments, current account (million US$)	−246	−706[b]
Tourist arrivals (in thousands)	405[c]	541[b]
GDP (million US$)	11180	9353[d]
GDP (per capita US$)	1404	988[d]
Long-term rate of change in GDP (% per annum)	−0.6	3.3[d]
Gross fixed capital formation (% of GDP)	11.0	12.1[d]
Economically active female population (%)	13.8[e]	17.8
Economically active male population (%)	85.4[e]	83.2
Annual growth of econ. active female pop. (%)	3.8[f]	...
Annual growth of econ. active male pop. (%)	2.5[f]	...
Labour force in industry (%)	16.9[e]	19.1[d]
Labour force in agriculture (%)	49.4[e]	30.2[d]
Agricultural production index (1979-1981=100)	109[c]	120[b]
Food production index (1979-1981=100)	123[c]	136[b]
Commercial energy production (1,000 Mt coal equiv)	324[e]	500[d]
Motor vehicles (per 1,000 population)	24	20[f]
Telephone lines (per 100 inhabitants)	1.6[g]	2.1[d]

Largest export industries	Major trading partners		1992
(% of exports)	(% of exports)		(% of imports)
...	United States	35	...
...	El Salvador	14	...
...	Costa Rica	7	...

Social indicators	1990/95
Growth rate of population (% per annum)	2.9
Age group 0-14 years (%)	44.3
Age group 60+ years (women and men, %)	2.8/2.6
Life expectancy at birth (women and men, years)	67/62
Infant mortality rate (per 1,000 births)	49
Total fertility rate (births per woman)	5
Contraceptive use (% of currently married women)	23[g]
Urban population (%)	41
Urban population growth rate (% per annum)	3.9
Rural population growth rate (% per annum)	2.2
Foreign-born (1985,%)	1.4
Refugees	222900
Government education expenditure (% of GDP)	2[h]
Primary-secondary gross enrolment ratio (f and m, per 100)	43/51[e]
Third-level students (per 100,000 population)	741[i]
Newspaper circulation (per 1,000 population)	21
Television receivers (per 1,000 population)	52
Intentional homicides (1986, per 100,000 population)	3[h,j]
Parliamentary seats (women and men, %)	5/95

Environment	1990/91
Threatened species	27
Forested area (%)	34.4
CO_2 emissions (10,000 Mt)	1112
Energy consumption (1,000 Mt coal equiv)	180
Precipitation (mm)	1281[g]
Average temperature (January and July, centigrade)	16.3/18.5[e]

a December 1993. b 1992. c 1988. d 1991. e 1980. f 1990. g 1987.
h 1984. i 1985. j Source: DYB92.

Guinea

Region	Western Africa
Location (longitude, latitude)	9°31'N 13°43'W
Currency	Guinean franc
Population (1995 est., in thousands)	6700
Surface area (square kms)	245857
Population density (pop. per square km)	24
Sex ratio (females per 100 males)	99
Largest city (pop. in thousands)	Conakry (1127)
UN membership date	12-Dec-1958
Major language(s)	French, Ful, Malinke, Susu, Kisi, Kpelle

Economic indicators	1985	1994
Exchange rate (US$)	22.47	975.97[a]
Consumer price index (1980=100)	195[b]	272
Balance of payments, current account (million US$)	−196[b]	−203[c]
GDP (million US$)	2508	3016[d]
GDP (per capita US$)	503	508[d]
Long-term rate of change in GDP (% per annum)	3.9	4.7
Economically active female population (%)	62.5[e]	54.9
Economically active male population (%)	91.1[e]	89.4
Annual growth of econ. active female pop. (%)	1.5[b]	...
Annual growth of econ. active male pop. (%)	2.0[b]	...
Agricultural production index (1979-1981=100)	115[f]	121[c]
Food production index (1979-1981=100)	114[f]	118[c]
Commercial energy production (1,000 Mt coal equiv)	18[e]	22[d]
Motor vehicles (per 1,000 population)	4	16[d]
Telephone lines (per 100 inhabitants)	0.4[g]	0.2[d]

Social indicators	1990/95
Growth rate of population (% per annum)	3.0
Age group 0-14 years (%)	47.1
Age group 60+ years (women and men, %)	2.3/2.0
Life expectancy at birth (women and men, years)	45/44
Infant mortality rate (per 1,000 births)	134
Total fertility rate (births per woman)	7
Urban population (%)	30
Urban population growth rate (% per annum)	5.8
Rural population growth rate (% per annum)	2.0
Foreign-born (1985,%)	0.3
Refugees	478500
Primary-secondary gross enrolment ratio (f and m, per 100)	15/33
Third-level students (per 100,000 population)	122[f]
Television receivers (per 1,000 population)	7

Environment	1990/91
Threatened species	27
Forested area (%)	59.3
CO_2 emissions (10,000 Mt)	280
Energy consumption (1,000 Mt coal equiv)	85

a June 1994. b 1990. c 1992. d 1991. e 1980. f 1988. g 1987.

Guinea-Bissau

Region	Western Africa
Location (longitude, latitude)	11°51'N 15°35'W[a]
Currency	Guinea-Bissau peso
Population (1995 est., in thousands)	1073
Surface area (square kms)	36125
Population density (pop. per square km)	27
Sex ratio (females per 100 males)	103
Largest city (pop. in thousands)	Bissau (71)[a]
UN membership date	17-Sep-1974
Major language(s)	English, Balante, Ful, Crioulo, Malinke, Mandyak

Economic indicators	1985	1994
Exchange rate (US$)	173.61	8655.60
Balance of payments, current account (million US$)	−45	−79[b]
GDP (million US$)	248	251[c]
GDP (per capita US$)	284	255[c]
Long-term rate of change in GDP (% per annum)	−2.3	3.0[c]
Gross fixed capital formation (% of GDP)	10.4[c]	...
Economically active female population (%)	62.7[d]	55.0
Economically active male population (%)	91.6[d]	89.7
Annual growth of econ. active female pop. (%)	2.4[e]	...
Annual growth of econ. active male pop. (%)	2.7[e]	...
Agricultural production index (1979-1981=100)	127[f]	138[b]
Food production index (1979-1981=100)	128[f]	139[b]
Motor vehicles (per 1,000 population)	8	...
Telephone lines (per 100 inhabitants)	0.6[g]	0.6[c]

Social indicators	1990/95
Growth rate of population (% per annum)	2.1
Age group 0-14 years (%)	41.7
Age group 60+ years (women and men, %)	3.4/3.0
Life expectancy at birth (women and men, years)	45/42
Infant mortality rate (per 1,000 births)	140
Total fertility rate (births per woman)	6
Urban population (%)	22
Urban population growth rate (% per annum)	4.4
Rural population growth rate (% per annum)	1.5
Foreign-born (1985,%)	1.7
Refugees	12200
Government education expenditure (% of GDP)	3[g]
Primary-secondary gross enrolment ratio (f and m, per 100)	27/49[f]
Newspaper circulation (per 1,000 population)	6
Parliamentary seats (women and men, %)	13/87

Environment	1990/91
Threatened species	11
Forested area (%)	29.6
CO_2 emissions (10,000 Mt)	56
Energy consumption (1,000 Mt coal equiv)	99
Precipitation (mm)	2023[g]
Average temperature (January and July, centigrade)	24.4/26.3[d]

a Capital city. b 1992. c 1991. d 1980. e 1990. f 1988. g 1987.

Guyana

Region	South America
Location (longitude, latitude)	6°82'N 58°15'W[a]
Currency	Guyana dollar
Population (1995 est., in thousands)	834
Surface area (square kms)	214969
Population density (pop. per square km)	4
Sex ratio (females per 100 males)	102
Largest city (pop. in thousands)	Georgetown (234)[a]
UN membership date	20-Sep-1966
Major language(s)	English, Hindi, Urdu

Economic indicators	1985	1994
Exchange rate (US$)	4.15	141.25
Consumer price index (1980=100)	245	...
Balance of payments, current account (million US$)	−97	...
Tourist arrivals (in thousands)	71[b]	93[c]
GDP (million US$)	462	219[d]
GDP (per capita US$)	585	274[d]
Long-term rate of change in GDP (% per annum)	1.0	6.0[d]
Gross fixed capital formation (% of GDP)	20.9	62.2[e]
Economically active female population (%)	26.3[f]	29.2
Economically active male population (%)	84.1[f]	85.0
Annual growth of econ. active female pop. (%)	4.2[e]	...
Annual growth of econ. active male pop. (%)	3.0[e]	...
Agricultural production index (1979-1981=100)	98[b]	100[c]
Food production index (1979-1981=100)	99[b]	101[c]
Commercial energy production (1,000 Mt coal equiv)	1[f]	1[d]
Motor vehicles (per 1,000 population)	52	41[e]
Telephone lines (per 100 inhabitants)	2.0[e]	2.0[d]

Social indicators	1990/95
Growth rate of population (% per annum)	0.9
Age group 0-14 years (%)	32.3
Age group 60+ years (women and men, %)	3.2/2.6
Life expectancy at birth (women and men, years)	68/62
Infant mortality rate (per 1,000 births)	48
Total fertility rate (births per woman)	3
Urban population (%)	35
Urban population growth rate (% per annum)	2.4
Rural population growth rate (% per annum)	0.2
Foreign-born (1985,%)	0.9
Government education expenditure (% of GDP)	4
Primary-secondary gross enrolment ratio (f and m, per 100)	83/82[b]
Third-level students (per 100,000 population)	587[g]
Newspaper circulation (per 1,000 population)	101
Television receivers (per 1,000 population)	39
Intentional homicides (1986, per 100,000 population)	0[h,i]

Environment	1990/91
Threatened species	22
Forested area (%)	76.1
CO_2 emissions (10,000 Mt)	232
Energy consumption (1,000 Mt coal equiv)	503
Precipitation (mm)	2418[j]
Average temperature (January and July, centigrade)	26.3/26.2[f]

a Capital city. b 1988. c 1992. d 1991. e 1990. f 1980. g 1989. h Source: DYB92. i 1984. j 1987.

Haiti

Region Caribbean
Location (longitude, latitude) 18°54'N 72°34'W
Currency gourde
Population (1995 est., in thousands) 7180
Surface area (square kms) 27750
Population density (pop. per square km) 239
Sex ratio (females per 100 males) 104
Largest city (pop. in thousands) Port-au-Prince (1041)
UN membership date 24-Oct-1945
Major language(s) Haitian, French

Economic indicators	1985	1994
Exchange rate (US$)	5.00[a]	15.00
Consumer price index (1980=100)	155	333[b]
Balance of payments, current account (million US$)	−95	−11[c]
Tourist arrivals (in thousands)	133[d]	120[e]
GDP (million US$)	2009	2951[c]
GDP (per capita US$)	343	446[c]
Long-term rate of change in GDP (% per annum)	0.6	0.7[c]
Gross fixed capital formation (% of GDP)	16.7	13.0[c]
Economically active female population (%)	60.6[f]	53.7
Economically active male population (%)	84.1[f]	83.4
Annual growth of econ. active female pop. (%)	1.1[g]	...
Annual growth of econ. active male pop. (%)	2.1[g]	...
Labour force in industry (%)	8.0[f]	8.9[d]
Labour force in agriculture (%)	67.5[f]	66.2[d]
Agricultural production index (1979-1981=100)	102[d]	80[e]
Food production index (1979-1981=100)	103[d]	81[e]
Commercial energy production (1,000 Mt coal equiv)	27[f]	40[c]
Motor vehicles (per 1,000 population)	6	...
Telephone lines (per 100 inhabitants)	0.7[g]	0.8[c]

Social indicators	1990/95
Growth rate of population (% per annum)	2.0
Age group 0-14 years (%)	40.2
Age group 60+ years (women and men, %)	3.3/2.8
Life expectancy at birth (women and men, years)	58/55
Infant mortality rate (per 1,000 births)	86
Total fertility rate (births per woman)	5
Contraceptive use (% of currently married women)	10[h]
Urban population (%)	32
Urban population growth rate (% per annum)	4.0
Rural population growth rate (% per annum)	1.2
Foreign-born (1985,%)	0.3
Government education expenditure (% of GDP)	2
Primary-secondary gross enrolment ratio (f and m, per 100)	39/41
Third-level students (per 100,000 population)	107[i]
Newspaper circulation (per 1,000 population)	7
Television receivers (per 1,000 population)	5
Parliamentary seats (women and men, %)	4/96

Environment	1990/91
Threatened species	28
Forested area (%)	1.4
CO_2 emissions (10,000 Mt)	200
Energy consumption (1,000 Mt coal equiv)	52

a Fixed rate. b September 1993. c 1991. d 1988. e 1992. f 1980. g 1990. h 1989. i 1985.

Honduras

Region	Central America
Location (longitude, latitude)	14°10'N 87°21'W
Currency	lempira
Population (1995 est., in thousands)	5968
Surface area (square kms)	112088
Population density (pop. per square km)	47
Sex ratio (females per 100 males)	98
Largest city (pop. in thousands)	Tegucigalpa (771)
UN membership date	17-Dec-1945
Major language(s)	Spanish

Economic indicators	1985	1994
Exchange rate (US$)	2.00[a]	8.77
Consumer price index (1980=100)	139	308
Industrial production index (1980 = 100)	116	140[b]
Balance of payments, current account (million US$)	–309	–264[c]
Tourist arrivals (in thousands)	162[d]	230[c]
GDP (million US$)	3504	2797[e]
GDP (per capita US$)	799	528[e]
Long-term rate of change in GDP (% per annum)	3.5	2.9[e]
Gross fixed capital formation (% of GDP)	17.9	...
Economically active female population (%)	17.4[f]	22.8
Economically active male population (%)	87.3[f]	85.8
Annual growth of econ. active female pop. (%)	5.2[b]	...
Annual growth of econ. active male pop. (%)	3.3[b]	...
Labour force in industry (%)	26.0[f]	41.5[c]
Labour force in agriculture (%)	1.0[f]	38.2[c]
Agricultural production index (1979-1981=100)	118[d]	145[c]
Food production index (1979-1981=100)	117[d]	134[c]
Commercial energy production (1,000 Mt coal equiv)	96[f]	109[e]
Motor vehicles (per 1,000 population)	8	15[b]
Telephone lines (per 100 inhabitants)	1.2[g]	1.8[e]

Largest export industries		Major trading partners			1992
	(% of exports)		(% of exports)		(% of imports)
Agriculture	75	United States	53	United States	53
Food, beverages, tobacco	9	Germany	13	Mexico	6
Textiles	3	Belgium[h]	5	Japan	5

Social indicators	1990/95
Growth rate of population (% per annum)	3.0
Age group 0-14 years (%)	43.2
Age group 60+ years (women and men, %)	2.6/2.4
Life expectancy at birth (women and men, years)	68/64
Infant mortality rate (per 1,000 births)	60
Total fertility rate (births per woman)	5
Contraceptive use (% of currently married women)	41[g]
Urban population (%)	48
Urban population growth rate (% per annum)	4.7
Rural population growth rate (% per annum)	1.5
Foreign-born (1985,%)	2.6
Refugees	100100
Government education expenditure (% of GDP)	4
Primary-secondary gross enrolment ratio (f and m, per 100)	76/71
Third-level students (per 100,000 population)	854
Newspaper circulation (per 1,000 population)	39
Television receivers (per 1,000 population)	73
Intentional homicides (1986, per 100,000 population)	8
Parliamentary seats (women and men, %)	12/88

Environment	1990/91
Threatened species	18
Forested area (%)	29.1
CO_2 emissions (10,000 Mt)	531
Energy consumption (1,000 Mt coal equiv)	166
Precipitation (mm)	954[g]
Average temperature (January and July, centigrade)	19.2/22.6[f]

a Fixed rate. b 1990. c 1992. d 1988. e 1991. f 1980. g 1987. h Includes Luxemburg.

Hong Kong

Region	Eastern Asia
Location (longitude, latitude)	22°18'N 114°10'E
Currency	Hong Kong dollar
Population (1995 est., in thousands)	5932
Surface area (square kms)	1045
Population density (pop. per square km)	5657
Sex ratio (females per 100 males)	94
Largest city (pop. in thousands)	Hong Kong (5374)
Major language(s)	Chinese, English

Economic indicators	1985	1994
Exchange rate (US$)	7.81	7.73
Consumer price index (1980=100)	155	319
Unemployment (%)	...	1.9[a]
Tourist arrivals (in thousands)	5589[b]	6986[c]
GDP (million US$)	33513	82837[d]
GDP (per capita US$)	6176	14396[d]
Long-term rate of change in GDP (% per annum)	0.2	4.2[d]
Gross fixed capital formation (% of GDP)	21.9	27.6[d]
Economically active female population (%)	46.3[e]	49.7
Economically active male population (%)	81.4[e]	80.1
Annual growth of econ. active female pop. (%)	3.2[f]	...
Annual growth of econ. active male pop. (%)	3.4[f]	...
Labour force in industry (%)	50.2[e]	32.9[c]
Labour force in agriculture (%)	1.4[e]	0.7[c]
Agricultural production index (1979-1981=100)	111[b]	99[c]
Food production index (1979-1981=100)	111[b]	99[c]
Motor vehicles (per 1,000 population)	44	63[d]
Telephone lines (per 100 inhabitants)	36.3[g]	45.9[d]

Largest export industries		Major trading partners			1992
	(% of exports)		(% of exports)		(% of imports)
Metal manufacture	36	China	30	China	37
Textiles	32	United States	23	Japan	17
Chemicals	10	Germany	5	United States	7

Social indicators	1990/95
Growth rate of population (% per annum)	0.8
Age group 0-14 years (%)	19.1
Age group 60+ years (women and men, %)	7.5/6.8
Life expectancy at birth (women and men, years)	80/75
Infant mortality rate (per 1,000 births)	6
Total fertility rate (births per woman)	1
Contraceptive use (% of currently married women)	81[g]
Urban population (%)	95
Urban population growth rate (% per annum)	1.0
Rural population growth rate (% per annum)	−2.6
Foreign-born (1985,%)	40.5
Refugees	45300
Government education expenditure (% of GDP)	2[h]
Primary-secondary gross enrolment ratio (f and m, per 100)	88/86[g]
Third-level students (per 100,000 population)	1201[e]
Newspaper circulation (per 1,000 population)	648
Television receivers (per 1,000 population)	278
Intentional homicides (1986, per 100,000 population)	1

Environment	1990/91
Threatened species	13
Forested area (%)	11.5
CO_2 emissions (10,000 Mt)	7951
Energy consumption (1,000 Mt coal equiv)	1920
Precipitation (mm)	2265[g]
Average temperature (January and July, centigrade)	15.4/28.4[e]

a Labour force sample surveys. b 1988. c 1992. d 1991. e 1980. f 1990.
g 1987. h 1989.

Hungary

Region	Eastern Europe
Location (longitude, latitude)	47°51'N 19°05'E
Currency	forint
Population (1995 est., in thousands)	10471
Surface area (square kms)	93032
Population density (pop. per square km)	111
Sex ratio (females per 100 males)	108
Largest city (pop. in thousands)	Budapest (2121)
UN membership date	14-Dec-1955
Major language(s)	Hungarian

Economic indicators	1985	1994
Exchange rate (US$)	47.35	107.88
Consumer price index (1980=100)	139	250
Industrial production index (1980 = 100)	110	87[a]
Balance of payments, current account (million US$)	−455	−4262[b]
Tourist arrivals (in thousands)	10563[c]	20188[d]
GDP (million US$)	21078	30888[e]
GDP (per capita US$)	1979	2933[e]
Long-term rate of change in GDP (% per annum)	−0.3	−11.9[e]
Gross fixed capital formation (% of GDP)	22.5	19.1[e]
Economically active female population (%)	50.0[f]	48.1
Economically active male population (%)	71.8[f]	68.6
Annual growth of econ. active female pop. (%)	0.4[g]	...
Annual growth of econ. active male pop. (%)	−0.6[g]	...
Labour force in industry (%)	41.1[f]	36.1[e]
Labour force in agriculture (%)	22.1[f]	16.1[e]
Agricultural production index (1979-1981=100)	116[c]	88[d]
Food production index (1979-1981=100)	117[c]	88[d]
Commercial energy production (1,000 Mt coal equiv)	21891[f]	20180[e]
Motor vehicles (per 1,000 population)	156	218[e]
Telephone lines (per 100 inhabitants)	7.7[h]	10.7[e]

Largest export industries	Major trading partners			1992
(% of exports)	(% of exports)		(% of imports)	
...	Germany	28	Germany	24
...	Russian Fed.	13	Russian Fed.	17
...	Austria	11	Austria	14

Social indicators	1990/95
Growth rate of population (% per annum)	−0.2
Age group 0-14 years (%)	18.5
Age group 60+ years (women and men, %)	11.6/7.6
Life expectancy at birth (women and men, years)	74/66
Infant mortality rate (per 1,000 births)	14
Total fertility rate (births per woman)	2
Contraceptive use (% of currently married women)	73[i]
Urban population (%)	68
Urban population growth rate (% per annum)	0.9
Rural population growth rate (% per annum)	−2.1
Foreign-born (1985, %)	0.2
Refugees	32400
Government education expenditure (% of GDP)	6
Primary-secondary gross enrolment ratio (f and m, per 100)	89/88
Third-level students (per 100,000 population)	970
Newspaper circulation (per 1,000 population)	233
Television receivers (per 1,000 population)	412
Intentional homicides (1986, per 100,000 population)	4[j]
Parliamentary seats (women and men, %)	7/93

Environment	1990/91
Threatened species	56
Forested area (%)	18.2
CO_2 emissions (10,000 Mt)	17351
Energy consumption (1,000 Mt coal equiv)	3574
Precipitation (mm)	630[h]
Average temperature (January and July, centigrade)	−1.1/22.2[f]

a June 1994. b 1993. c 1988. d 1992. e 1991. f 1980. g 1990. h 1987.
i 1986. j Source: DYB92.

Iceland

Region	Northern Europe
Location (longitude, latitude)	64°14'N 21°94'W[a]
Currency	Icelandic kronur
Population (1995 est., in thousands)	268
Surface area (square kms)	103000
Population density (pop. per square km)	3
Sex ratio (females per 100 males)	100
Largest city (pop. in thousands)	Reykjavik (146)[a]
UN membership date	19-Nov-1946
Major language(s)	Icelandic

Economic indicators	1985	1994
Exchange rate (US$)	42.06	67.71
Consumer price index (1980=100)	715	2105[b]
Unemployment (%)	...	3.5
Balance of payments, current account (million US$)	-115	-210[c]
Tourist arrivals (in thousands)	129[d]	143[c]
GDP (million US$)	2871	6490[e]
GDP (per capita US$)	11914	25252[e]
Long-term rate of change in GDP (% per annum)	3.9	1.4[e]
Gross fixed capital formation (% of GDP)	20.5	19.0[e]
Economically active female population (%)	59.1[f]	60.8
Economically active male population (%)	83.4[f]	81.7
Annual growth of econ. active female pop. (%)	3.2[g]	...
Annual growth of econ. active male pop. (%)	1.5[g]	...
Labour force in industry (%)	35.4[f]	29.5[g]
Labour force in agriculture (%)	13.2[f]	10.5[g]
Agricultural production index (1979-1981=100)	86[d]	90[c]
Food production index (1979-1981=100)	86[d]	90[c]
Commercial energy production (1,000 Mt coal equiv)	437[f]	863[e]
Motor vehicles (per 1,000 population)	486	537[e]
Telephone lines (per 100 inhabitants)	45.8[h]	52.7[e]

Largest export industries		Major trading partners			1992
	(% of exports)		(% of exports)		(% of imports)
Agriculture	74	United Kingdom	25	Norway[i]	15
Basic metal industry	11	Germany	13	Germany	12
Food, beverages, tobacco	10	United States	11	Denmark	9

Social indicators	1990/95
Growth rate of population (% per annum)	1.0
Age group 0-14 years (%)	24.3
Age group 60+ years (women and men, %)	7.8/7.1
Life expectancy at birth (women and men, years)	81/76
Infant mortality rate (per 1,000 births)	5
Total fertility rate (births per woman)	2
Urban population (%)	92
Urban population growth rate (% per annum)	1.3
Rural population growth rate (% per annum)	-1.2
Foreign-born (1985, %)	2.7
Refugees	200
Government education expenditure (% of GDP)	5
Third-level students (per 100,000 population)	2154[i]
Newspaper circulation (per 1,000 population)	467[f]
Television receivers (per 1,000 population)	319
Intentional homicides (1986, per 100,000 population)	2[k]
Parliamentary seats (women and men, %)	24/76

Environment	1990/91
Threatened species	5
Forested area (%)	1.2
CO_2 emissions (10,000 Mt)	492
Energy consumption (1,000 Mt coal equiv)	6498
Precipitation (mm)	805[h]
Average temperature (January and July, centigrade)	-0.4/11.2[f]

a Capital city. b August 1994. c 1992. d 1988. e 1991. f 1980. g 1990. h 1987. i Includes Svalbard and Jan Mayen Islands. j 1989. k Source: DYB92.

India

Region	Southern Asia
Location (longitude, latitude)	18°95'N 72°66'E
Currency	Indian rupee
Population (1995 est., in thousands)	931044
Surface area (square kms)	3287590
Population density (pop. per square km)	258
Sex ratio (females per 100 males)	94
Largest city (pop. in thousands)	Bombay (12223)
UN membership date	30-Oct-1945
Major language(s)	Hindi, English, Bengali, Telugu, Marathi, Tamil

Economic indicators	1985	1994
Exchange rate (US$)	12.16	31.37
Consumer price index (1980=100)	156	359[a]
Industrial production index (1980 = 100)	140	225[b]
Balance of payments, current account (million US$)	−4177	−7037[c]
Tourist arrivals (in thousands)	1591[d]	1868[e]
GDP (million US$)	212016	268006[f]
GDP (per capita US$)	277	311[f]
Long-term rate of change in GDP (% per annum)	5.5	1.3[f]
Gross fixed capital formation (% of GDP)	20.7	23.1[c]
Economically active female population (%)	32.2[g]	28.5
Economically active male population (%)	84.8[g]	84.0
Annual growth of econ. active female pop. (%)	1.2[c]	...
Annual growth of econ. active male pop. (%)	2.3[c]	...
Labour force in industry (%)	38.7[g]	36.3[h]
Labour force in agriculture (%)	5.8[g]	5.6[h]
Agricultural production index (1979-1981=100)	138[d]	155[e]
Food production index (1979-1981=100)	140[d]	157[e]
Commercial energy production (1,000 Mt coal equiv)	114066[g]	264714[f]
Motor vehicles (per 1,000 population)	5	8[f]
Telephone lines (per 100 inhabitants)	0.5[i]	0.7[f]

Social indicators	1990/95
Growth rate of population (% per annum)	1.9
Age group 0-14 years (%)	34.9
Age group 60+ years (women and men, %)	3.8/3.8
Life expectancy at birth (women and men, years)	61/60
Infant mortality rate (per 1,000 births)	88
Total fertility rate (births per woman)	4
Contraceptive use (% of currently married women)	43[d]
Urban population (%)	27
Urban population growth rate (% per annum)	2.9
Rural population growth rate (% per annum)	1.6
Foreign-born (1985,%)	1.2
Refugees	258400
Government education expenditure (% of GDP)	3[i]
Primary-secondary gross enrolment ratio (f and m, per 100)	55/79
Third-level students (per 100,000 population)	581[j]
Newspaper circulation (per 1,000 population)	26[g]
Television receivers (per 1,000 population)	35
Intentional homicides (1986, per 100,000 population)	3
Parliamentary seats (women and men, %)	7/93

Environment	1990/91
Threatened species	172
Forested area (%)	20.3
CO_2 emissions (10,000 Mt)	192017
Energy consumption (1,000 Mt coal equiv)	336
Precipitation (mm)	715[k]
Average temperature (January and July, centigrade)	14.3/31.2[g]

a August 1994. b April 1994. c 1990. d 1988. e 1992. f 1991. g 1980.
h 1989. i 1987. j 1985. k New Delhi.

Indonesia

Region South-eastern Asia
Location (longitude, latitude) 6°18'S 106°89'E
Currency rupiah
Population (1995 est., in thousands) 201477
Surface area (square kms) 1904569
Population density (pop. per square km) 99
Sex ratio (females per 100 males) 101
Largest city (pop. in thousands) Jakarta (9206)
UN membership date 28-Sep-1950
Major language(s) Indonesian, Javanese, Sudanese

Economic indicators	1985	1994
Exchange rate (US$)	1125.00	2181.00
Consumer price index (1980=100)	159	156[a]
Balance of payments, current account (million US$)	−1923	−3679[b]
Tourist arrivals (in thousands)	1301[c]	3064[b]
GDP (million US$)	87339	116476[d]
GDP (per capita US$)	522	620[d]
Long-term rate of change in GDP (% per annum)	2.5	6.6
Gross fixed capital formation (% of GDP)	23.1	26.9[d]
Economically active female population (%)	36.9[e]	36.6
Economically active male population (%)	84.7[e]	83.0
Annual growth of econ. active female pop. (%)	2.6[f]	...
Annual growth of econ. active male pop. (%)	2.3[f]	...
Labour force in industry (%)	13.1[e]	14.1[b]
Labour force in agriculture (%)	56.3[e]	54.9[b]
Agricultural production index (1979-1981=100)	147[c]	177[b]
Food production index (1979-1981=100)	150[c]	182[b]
Commercial energy production (1,000 Mt coal equiv)	132256[e]	180506[d]
Motor vehicles (per 1,000 population)	12	17[d]
Telephone lines (per 100 inhabitants)	0.4[g]	0.7[d]

Largest export industries		Major trading partners			1992
	(% of exports)		(% of exports)		(% of imports)
Mining, quarrying	32	Japan	32	Japan	22
Textiles	22	United States	13	United States	14
Wood, wood products	12	Singapore	9	Germany	8

Social indicators	1990/95
Growth rate of population (% per annum)	1.8
Age group 0-14 years (%)	33.4
Age group 60+ years (women and men, %)	3.7/3.3
Life expectancy at birth (women and men, years)	65/61
Infant mortality rate (per 1,000 births)	65
Total fertility rate (births per woman)	3
Contraceptive use (% of currently married women)	50
Urban population (%)	33
Urban population growth rate (% per annum)	4.2
Rural population growth rate (% per annum)	0.7
Foreign-born (1985,%)	0.7
Refugees	15600
Government education expenditure (% of GDP)	1[c]
Primary-secondary gross enrolment ratio (f and m, per 100)	78/85[h]
Third-level students (per 100,000 population)	375[i]
Newspaper circulation (per 1,000 population)	28
Television receivers (per 1,000 population)	59
Intentional homicides (1986, per 100,000 population)	1
Parliamentary seats (women and men, %)	12/88

Environment	1990/91
Threatened species	354
Forested area (%)	59.6
CO_2 emissions (10,000 Mt)	46525[j]
Energy consumption (1,000 Mt coal equiv)	348
Precipitation (mm)	1755[g]
Average temperature (January and July, centigrade)	26.2/26.7[e]

a June 1994. b 1992. c 1988. d 1991. e 1980. f 1990. g 1987. h 1989.
i 1981. j Includes East Timor.

Iran, Islamic Republic of

Region Southern Asia
Location (longitude, latitude) 35°69'N 51°41'E
Currency Iranian rial
Population (1995 est., in thousands) 66720
Surface area (square kms) 1648000
Population density (pop. per square km) 34
Sex ratio (females per 100 males) 97
Largest city (pop. in thousands) Teheran (6654)
UN membership date 24-Oct-1945
Major language(s) Persian, Azerbaijani, Kurdish

Economic indicators	1985	1994
Exchange rate (US$)	84.23	67.04
Consumer price index (1980=100)	84	162[a]
Balance of payments, current account (million US$)	−476	−7909[b]
Tourist arrivals (in thousands)	67[c]	185[d]
GDP (million US$)	173253	649018[b]
GDP (per capita US$)	3542	10826[b]
Long-term rate of change in GDP (% per annum)	1.8	4.6[b]
Gross fixed capital formation (% of GDP)	17.5	15.5[e]
Economically active female population (%)	14.5[f]	18.6
Economically active male population (%)	81.0[f]	78.1
Annual growth of econ. active female pop. (%)	5.2[e]	...
Annual growth of econ. active male pop. (%)	3.2[e]	...
Agricultural production index (1979-1981=100)	144[c]	189[d]
Food production index (1979-1981=100)	144[c]	192[d]
Commercial energy production (1,000 Mt coal equiv)	116046[f]	284345[b]
Motor vehicles (per 1,000 population)	53	
Telephone lines (per 100 inhabitants)	2.8[g]	4.1[b]

Social indicators	1990/95
Growth rate of population (% per annum)	2.7
Age group 0-14 years (%)	45.9
Age group 60+ years (women and men, %)	2.9/3.0
Life expectancy at birth (women and men, years)	68/67
Infant mortality rate (per 1,000 births)	40
Total fertility rate (births per woman)	6
Contraceptive use (% of currently married women)	65
Urban population (%)	60
Urban population growth rate (% per annum)	3.9
Rural population growth rate (% per annum)	1.0
Foreign-born (1985,%)	5.9
Refugees	4150700
Government education expenditure (% of GDP)	4
Primary-secondary gross enrolment ratio (f and m, per 100)	75/90
Third-level students (per 100,000 population)	1061
Newspaper circulation (per 1,000 population)	26
Television receivers (per 1,000 population)	63
Parliamentary seats (women and men, %)	3/97

Environment	1990/91
Threatened species	52
Forested area (%)	10.9
CO_2 emissions (10,000 Mt)	60688
Energy consumption (1,000 Mt coal equiv)	1654
Precipitation (mm)	208[g]
Average temperature (January and July, centigrade)	3.5/29.5[f]

a March 1994. b 1991. c 1988. d 1992. e 1990. f 1980. g 1987.

Iraq

Region	Western Asia
Location (longitude, latitude)	33°34'N 44°43'E
Currency	Iraqi dinar
Population (1995 est., in thousands)	21224
Surface area (square kms)	438317
Population density (pop. per square km)	45
Sex ratio (females per 100 males)	96
Largest city (pop. in thousands)	Baghdad (4044)
UN membership date	21-Dec-1945
Major language(s)	Arabic, Kurdish

Economic indicators	1985	1994
Exchange rate (US$)	0.31	0.31
Consumer price index (1980=100)	199	161[a]
Tourist arrivals (in thousands)	1209[b]	504[c]
GDP (million US$)	49819	62628[d]
GDP (per capita US$)	3252	3353[d]
Long-term rate of change in GDP (% per annum)	0.0	−46.0[d]
Gross fixed capital formation (% of GDP)	27.8	...
Economically active female population (%)	17.3[e]	22.8
Economically active male population (%)	78.9[e]	77.4
Annual growth of econ. active female pop. (%)	10.3[a]	...
Annual growth of econ. active male pop. (%)	3.3[a]	...
Agricultural production index (1979-1981=100)	127[b]	122[c]
Food production index (1979-1981=100)	129[b]	123[c]
Commercial energy production (1,000 Mt coal equiv)	188237[e]	20685[d]
Motor vehicles (per 1,000 population)	35	58[a]
Telephone lines (per 100 inhabitants)	4.1[f]	3.6[d]

Social indicators	1990/95
Growth rate of population (% per annum)	3.2
Age group 0-14 years (%)	43.8
Age group 60+ years (women and men, %)	2.4/2.2
Life expectancy at birth (women and men, years)	67/65
Infant mortality rate (per 1,000 births)	58
Total fertility rate (births per woman)	6
Contraceptive use (% of currently married women)	14[g]
Urban population (%)	75
Urban population growth rate (% per annum)	4.0
Rural population growth rate (% per annum)	1.1
Foreign-born (1985,%)	3.3
Refugees	95000
Primary-secondary gross enrolment ratio (f and m, per 100)	72/92
Third-level students (per 100,000 population)	1188[b]
Newspaper circulation (per 1,000 population)	36
Television receivers (per 1,000 population)	72
Parliamentary seats (women and men, %)	11/89

Environment	1990/91
Threatened species	30
Forested area (%)	4.3
CO_2 emissions (10,000 Mt)	11560
Energy consumption (1,000 Mt coal equiv)	970
Precipitation (mm)	151[f]
Average temperature (January and July, centigrade)	9.9/34.8[e]

a 1990. b 1988. c 1992. d 1991. e 1980. f 1987. g 1989.

Ireland

Region	Northern Europe
Location (longitude, latitude)	53°34'N 6°25'W
Currency	Irish pound
Population (1995 est., in thousands)	3469
Surface area (square kms)	70284
Population density (pop. per square km)	50
Sex ratio (females per 100 males)	100
Largest city (pop. in thousands)	Dublin (926)
UN membership date	14-Dec-1955
Major language(s)	Irish, English

Economic indicators	1985	1994
Exchange rate (US$)	0.80	0.64
Consumer price index (1980=100)	178	233[a]
Industrial production index (1980 = 100)	128	261[b]
Unemployment (%)	...	15.0[c]
Balance of payments, current account (million US$)	−690	2629[d]
Tourist arrivals (in thousands)	3007[e]	3666[d]
GDP (million US$)	18726	43593[f]
GDP (per capita US$)	5272	12480[f]
Long-term rate of change in GDP (% per annum)	3.1	2.5[f]
Gross fixed capital formation (% of GDP)	19.0	18.6[g]
Economically active female population (%)	29.7[h]	30.7
Economically active male population (%)	76.8[h]	75.7
Annual growth of econ. active female pop. (%)	2.0[g]	...
Annual growth of econ. active male pop. (%)	1.2[g]	...
Labour force in industry (%)	29.4[i]	28.6[f]
Labour force in agriculture (%)	16.9[i]	13.7[f]
Agricultural production index (1979-1981=100)	114[e]	132[d]
Food production index (1979-1981=100)	114[e]	132[d]
Commercial energy production (1,000 Mt coal equiv)	2723[h]	4733[f]
Motor vehicles (per 1,000 population)	229	272[g]
Telephone lines (per 100 inhabitants)	22.4[i]	30.0[f]

Largest export industries		Major trading partners			1992
	(% of exports)		(% of exports)		(% of imports)
Metal manufacture	39	United Kingdom	32	United Kingdom	42
Chemicals	23	Germany	13	United States	14
Food, beverages, tobacco	23	France[k]	10	Germany	8

Social indicators	1990/95
Growth rate of population (% per annum)	−0.2
Age group 0-14 years (%)	24.7
Age group 60+ years (women and men, %)	8.6/6.9
Life expectancy at birth (women and men, years)	78/73
Infant mortality rate (per 1,000 births)	7
Total fertility rate (births per woman)	2
Urban population (%)	58
Urban population growth rate (% per annum)	0.3
Rural population growth rate (% per annum)	−0.8
Foreign-born (1985,%)	8.0
Refugees	500
Government education expenditure (% of GDP)	6
Primary-secondary gross enrolment ratio (f and m, per 100)	101/97[l]
Third-level students (per 100,000 population)	2308[l]
Newspaper circulation (per 1,000 population)	169
Television receivers (per 1,000 population)	76
Intentional homicides (1986, per 100,000 population)	1
Parliamentary seats (women and men, %)	12/88

Environment	1990/91
Threatened species	11
Forested area (%)	4.9
CO_2 emissions (10,000 Mt)	8798
Energy consumption (1,000 Mt coal equiv)	4143
Precipitation (mm)	758[j]
Average temperature (January and July, centigrade)	4.7/15.1[h]

a August 1994. b June 1994. c Unemployment insurance statistics. d 1992. e 1988. f 1991. g 1990. h 1980. i 1983. j 1987. k Includes Monaco. l 1989.

Israel

Region	Western Asia
Location (longitude, latitude)	32°07'N 34°79'E
Currency	new shekel
Population (1995 est., in thousands)	5884
Surface area (square kms)	21056
Population density (pop. per square km)	236
Sex ratio (females per 100 males)	102
Largest city (pop. in thousands)	Tel-Aviv (1902)
UN membership date	11-May-1949
Major language(s)	Hebrew, Arabic

Economic indicators	1985	1994
Exchange rate (US$)	1.50	3.01
Consumer price index (1980=100)	22500	111772
Industrial production index (1980 = 100)	120	184[a]
Unemployment (%)	...	7.4[b]
Balance of payments, current account (million US$)	1144	−1373[c]
Tourist arrivals (in thousands)	1170[d]	1502[e]
GDP (million US$)	25867	62684[f]
GDP (per capita US$)	6111	12869[f]
Long-term rate of change in GDP (% per annum)	3.6	7.1[f]
Gross fixed capital formation (% of GDP)	17.5	22.4[f]
Economically active female population (%)	36.7[g]	36.9
Economically active male population (%)	75.3[g]	75.1
Annual growth of econ. active female pop. (%)	3.2[h]	...
Annual growth of econ. active male pop. (%)	2.3[h]	...
Labour force in industry (%)	30.8[g]	28.5[e]
Labour force in agriculture (%)	6.3[g]	3.5[e]
Agricultural production index (1979-1981=100)	114[d]	110[e]
Food production index (1979-1981=100)	121[d]	126[e]
Commercial energy production (1,000 Mt coal equiv)	220[g]	50[f]
Motor vehicles (per 1,000 population)	172	209[h]
Telephone lines (per 100 inhabitants)	32.1[i]	34.9[f]

Largest export industries		Major trading partners			1992
	(% of exports)		(% of exports)		(% of imports)
Metal manufacture	33	United States	31	United States	17
Mining, quarrying	30	United Kingdom	8	Belgium[j]	13
Chemicals	16	Germany	6	Germany	12

Social indicators	1990/95
Growth rate of population (% per annum)	4.7
Age group 0-14 years (%)	28.8
Age group 60+ years (women and men, %)	7.4/5.7
Life expectancy at birth (women and men, years)	78/75
Infant mortality rate (per 1,000 births)	9
Total fertility rate (births per woman)	3
Urban population (%)	93
Urban population growth rate (% per annum)	4.9
Rural population growth rate (% per annum)	2.0
Foreign-born (1985, %)	33.9
Government education expenditure (% of GDP)	6[k]
Primary-secondary gross enrolment ratio (f and m, per 100)	92/88[k]
Third-level students (per 100,000 population)	2655[k]
Newspaper circulation (per 1,000 population)	258
Television receivers (per 1,000 population)	269
Intentional homicides (1986, per 100,000 population)	6
Parliamentary seats (women and men, %)	9/91

Environment	1990/91
Threatened species	30
Forested area (%)	5.4
CO_2 emissions (10,000 Mt)	9707
Energy consumption (1,000 Mt coal equiv)	2999
Precipitation (mm)	492[l]
Average temperature (January and July, centigrade)	8.6/23.3[g]

a June 1994. b Labour force sample surveys. c 1993. d 1988. e 1992.
f 1991. g 1980. h 1990. i 1987. j Includes Luxemburg. k 1989.
l Jerusalem.

Italy

Region	Southern Europe
Location (longitude, latitude)	45°45'N 9°18'E
Currency	Italian lire
Population (1995 est., in thousands)	57910
Surface area (square kms)	301268
Population density (pop. per square km)	189
Sex ratio (females per 100 males)	106
Largest city (pop. in thousands)	Milan (5279)
UN membership date	14-Dec-1955
Major language(s)	Italian

Economic indicators	1985	1994
Exchange rate (US$)	1678.50	1556.50
Consumer price index (1980=100)	190	305[a]
Industrial production index (1980 = 100)	97	123[a]
Unemployment (%)	...	11.6[b]
Balance of payments, current account (million US$)	–3865	–27994[c]
Tourist arrivals (in thousands)	26155[d]	26113[c]
GDP (million US$)	424512	1150526[e]
GDP (per capita US$)	7429	19930[e]
Long-term rate of change in GDP (% per annum)	2.6	1.4
Gross fixed capital formation (% of GDP)	20.7	19.8[e]
Economically active female population (%)	29.9[f]	30.2
Economically active male population (%)	70.1[f]	68.6
Annual growth of econ. active female pop. (%)	1.1[g]	...
Annual growth of econ. active male pop. (%)	0.3[g]	...
Labour force in industry (%)	37.2[f]	32.0[e]
Labour force in agriculture (%)	14.0[f]	8.4[e]
Agricultural production index (1979-1981=100)	100[d]	105[c]
Food production index (1979-1981=100)	100[d]	105[c]
Commercial energy production (1,000 Mt coal equiv)	29246[f]	39774[e]
Motor vehicles (per 1,000 population)	434	538[e]
Telephone lines (per 100 inhabitants)	33.3[h]	40.0[e]

Largest export industries		Major trading partners			1992
	(% of exports)		(% of exports)		(% of imports)
Metal manufacture	47	Germany	20	Germany	22
Textiles	18	France[i]	15	France[i]	15
Chemicals	12	United States	7	Netherlands	6

Social indicators	1990/95
Growth rate of population (% per annum)	0.9
Age group 0-14 years (%)	15.4
Age group 60+ years (women and men, %)	12.3/9.0
Life expectancy at birth (women and men, years)	80/74
Infant mortality rate (per 1,000 births)	8
Total fertility rate (births per woman)	1
Urban population (%)	71
Urban population growth rate (% per annum)	0.6
Rural population growth rate (% per annum)	–1.0
Foreign-born (1985,%)	2.3
Refugees	12400
Government education expenditure (% of GDP)	3
Primary-secondary gross enrolment ratio (f and m, per 100)	84/83[j]
Third-level students (per 100,000 population)	2545
Newspaper circulation (per 1,000 population)	106[j]
Television receivers (per 1,000 population)	421
Intentional homicides (1986, per 100,000 population)	4
Parliamentary seats (women and men, %)	8/92

Environment	1990/91
Threatened species	76
Forested area (%)	22.4
CO_2 emissions (10,000 Mt)	109857[i]
Energy consumption (1,000 Mt coal equiv)	3998
Precipitation (mm)	903[h]
Average temperature (January and July, centigrade)	0.6/23.0[f]

a July 1994. b Labour force sample surveys. c 1992. d 1988. e 1991.
f 1980. g 1990. h 1987. i Includes Monaco. j 1989.

Jamaica

Region	Caribbean
Location (longitude, latitude)	18°00'N 76°77'W[a]
Currency	Jamaican dollar
Population (1995 est., in thousands)	2547
Surface area (square kms)	10990
Population density (pop. per square km)	215
Sex ratio (females per 100 males)	100
Largest city (pop. in thousands)	Kingston (638)[a]
UN membership date	18-Sep-1962
Major language(s)	English

Economic indicators	1985	1994
Exchange rate (US$)	5.48	33.28
Consumer price index (1980=100)	214	546[b]
Balance of payments, current account (million US$)	–304	117[c]
Tourist arrivals (in thousands)	649[d]	909[c]
GDP (million US$)	2015	3497[e]
GDP (per capita US$)	872	1431[e]
Long-term rate of change in GDP (% per annum)	–4.6	0.2[e]
Gross fixed capital formation (% of GDP)	23.0	29.1
Economically active female population (%)	65.4[g]	68.3
Economically active male population (%)	83.0[g]	83.1
Annual growth of econ. active female pop. (%)	3.3[f]	...
Annual growth of econ. active male pop. (%)	2.6[f]	...
Labour force in industry (%)	15.2[g]	22.6[f]
Labour force in agriculture (%)	37.3[g]	26.1[f]
Agricultural production index (1979-1981=100)	106[d]	126[c]
Food production index (1979-1981=100)	106[d]	126[c]
Commercial energy production (1,000 Mt coal equiv)	15[g]	16[e]
Motor vehicles (per 1,000 population)	30	44[e]
Telephone lines (per 100 inhabitants)	3.5[h]	4.7[e]

Largest export industries	Major trading partners		1992
(% of exports)	(% of exports)		(% of imports)
...	United States	37	United States 53
...	United Kingdom	17	Mexico 7
...	Canada	11	Japan 5

Social indicators	1990/95
Growth rate of population (% per annum)	1.0
Age group 0-14 years (%)	30.9
Age group 60+ years (women and men, %)	4.8/4.0
Life expectancy at birth (women and men, years)	76/71
Infant mortality rate (per 1,000 births)	14
Total fertility rate (births per woman)	2
Contraceptive use (% of currently married women)	67
Urban population (%)	55
Urban population growth rate (% per annum)	2.2
Rural population growth rate (% per annum)	–0.3
Foreign-born (1985,%)	0.9
Government education expenditure (% of GDP)	4
Primary-secondary gross enrolment ratio (f and m, per 100)	82/78[i]
Third-level students (per 100,000 population)	515[i]
Newspaper circulation (per 1,000 population)	64
Television receivers (per 1,000 population)	131
Intentional homicides (1986, per 100,000 population)	19
Parliamentary seats (women and men, %)	12/88

Environment	1990/91
Threatened species	29
Forested area (%)	16.8
CO_2 emissions (10,000 Mt)	1275
Energy consumption (1,000 Mt coal equiv)	834
Precipitation (mm)	811[h]
Average temperature (January and July, centigrade)	25.4/28.3[g]

a Capital city. b December 1993. c 1992. d 1988. e 1991. f 1990.
g 1980. h 1987. i 1989.

Japan

Region Eastern Asia
Location (longitude, latitude) 35°68'N 139°77'E
Currency yen
Population (1995 est., in thousands) 125879
Surface area (square kms) 377801
Population density (pop. per square km) 328
Sex ratio (females per 100 males) 103
Largest city (pop. in thousands) Tokyo (25013)
UN membership date 18-Dec-1956
Major language(s) Japanese

Economic indicators	1985	1994
Exchange rate (US$)	200.50	98.45
Consumer price index (1980=100)	114	131
Industrial production index (1980 = 100)	118	143[a]
Unemployment (%)	...	3.0[b]
Balance of payments, current account (million US$)	49	131[c]
Tourist arrivals (in thousands)	1116[d]	2103[e]
GDP (million US$)	1343251	3346411[f]
GDP (per capita US$)	11116	26983[f]
Long-term rate of change in GDP (% per annum)	5.0	4.1[f]
Gross fixed capital formation (% of GDP)	27.5	31.7[f]
Economically active female population (%)	47.3[g]	49.7
Economically active male population (%)	82.0[g]	78.3
Annual growth of econ. active female pop. (%)	0.6[h]	...
Annual growth of econ. active male pop. (%)	0.9[h]	...
Labour force in industry (%)	35.4[g]	34.6[e]
Labour force in agriculture (%)	10.4[g]	6.4[e]
Agricultural production index (1979-1981=100)	97[d]	96[e]
Food production index (1979-1981=100)	100[d]	99[e]
Commercial energy production (1,000 Mt coal equiv)	62939[g]	104816[f]
Motor vehicles (per 1,000 population)	374	475[f]
Telephone lines (per 100 inhabitants)	39.7[i]	45.4[f]

Largest export industries		Major trading partners			1992
	(% of exports)		(% of exports)		(% of imports)
Metal manufacture	78	United States	28	United States	23
Chemicals	9	Hong Kong	6	China	7
Basic metal industry	5	Germany	6	Indonesia	5

Social indicators	1990/95
Growth rate of population (% per annum)	0.4
Age group 0-14 years (%)	16.8
Age group 60+ years (women and men, %)	11.1/8.76
Life expectancy at birth (women and men, years)	82/76
Infant mortality rate (per 1,000 births)	5
Total fertility rate (births per woman)	2
Contraceptive use (% of currently married women)	64
Urban population (%)	78
Urban population growth rate (% per annum)	0.6
Rural population growth rate (% per annum)	−0.3
Foreign-born (1985,%)	0.6
Refugees	8200
Government education expenditure (% of GDP)	5[d]
Primary-secondary gross enrolment ratio (f and m, per 100)	99/98[j]
Third-level students (per 100,000 population)	2184[j]
Newspaper circulation (per 1,000 population)	587
Television receivers (per 1,000 population)	613
Intentional homicides (1986, per 100,000 population)	1
Parliamentary seats (women and men, %)	2/98

Environment	1990/91
Threatened species	151
Forested area (%)	66.5
CO_2 emissions (10,000 Mt)	297802
Energy consumption (1,000 Mt coal equiv)	4754
Precipitation (mm)	1563[i]
Average temperature (January and July, centigrade)	3.7/25.1[g]

a August 1994. b Labour force sample surveys. c 1993. d 1988. e 1992.
f 1991. g 1980. h 1990. i 1987. j 1989.

Jordan

Region Western Asia
Location (longitude, latitude) 31°95'N 35°94'E
Currency Jordanian dinar
Population (1995 est., in thousands) 4755
Surface area (square kms) 97740
Population density (pop. per square km) 42
Sex ratio (females per 100 males) 95
Largest city (pop. in thousands) Amman (955)
UN membership date 14-Dec-1955
Major language(s) Arabic

Economic indicators	1985	1994
Exchange rate (US$)	0.37	0.70
Consumer price index (1980=100)	130	241[a]
Industrial production index (1980 = 100)	155	178[b]
Balance of payments, current account (million US$)	−261	−765[b]
Tourist arrivals (in thousands)	608[c]	661[b]
GDP (million US$)	4818	4041[d]
GDP (per capita US$)	1414	974[d]
Long-term rate of change in GDP (% per annum)	3.5	0.5[d]
Gross fixed capital formation (% of GDP)	24.0	20.0[e]
Economically active female population (%)	6.8[f]	10.3
Economically active male population (%)	76.1[f]	77.8
Annual growth of econ. active female pop. (%)	5.2[e]	...
Annual growth of econ. active male pop. (%)	2.6[e]	...
Labour force in industry (%)	21.0[f]	22.8[d]
Agricultural production index (1979-1981=100)	167[c]	214[b]
Food production index (1979-1981=100)	169[c]	217[b]
Commercial energy production (1,000 Mt coal equiv)	25[e]	11[d]
Motor vehicles (per 1,000 population)	58	57[d]
Telephone lines (per 100 inhabitants)	5.6[g]	6.4[d]

Largest export industries		Major trading partners			1992
	(% of exports)		(% of exports)		(% of imports)
Chemicals	30	India	12	Iraq	13
Mining, quarrying	26	Saudi Arabia	9	United States	11
Metal manufacture	15	Iraq	9	Germany	9

Social indicators	1990/95
Growth rate of population (% per annum)	3.4
Age group 0-14 years (%)	43.6
Age group 60+ years (women and men, %)	2.3/2.1
Life expectancy at birth (women and men, years)	70/66
Infant mortality rate (per 1,000 births)	36
Total fertility rate (births per woman)	6
Contraceptive use (% of currently married women)	35
Urban population (%)	71
Urban population growth rate (% per annum)	4.4
Rural population growth rate (% per annum)	1.1
Foreign-born (1985,%)	26.5
Refugees	300
Government education expenditure (% of GDP)	4
Primary-secondary gross enrolment ratio (f and m, per 100)	93/91[h]
Third-level students (per 100,000 population)	2006
Newspaper circulation (per 1,000 population)	56
Television receivers (per 1,000 population)	80
Intentional homicides (1986, per 100,000 population)	2
Parliamentary seats (women and men, %)	0/100

Environment	1990/91
Threatened species	16
Forested area (%)	0.8
CO_2 emissions (10,000 Mt)	2732
Energy consumption (1,000 Mt coal equiv)	1030
Precipitation (mm)	273[g]
Average temperature (January and July, centigrade)	8.2/25.2[f]

a August 1994. b 1992. c 1988. d 1991. e 1990. f 1980. g 1987. h 1989.

Kazakhstan

Region	Central Asia
Location (longitude, latitude)	43°14'N 76°56'E
Currency	rouble
Surface area (square kms)	2717300
Largest city (pop. in thousands)	Alma-Ata (1158)
UN membership date	02-Mar-1992
Major language(s)	Kazakh, Russian

Economic indicators	1985	1994
Exchange rate (US$)	...	12.00
GDP (million US$)	37103	46445[a]
GDP (per capita US$)	2344	2749[a]
Long-term rate of change in GDP (% per annum)	6.2	−8.6[a]
Economically active female population (%)	54.2[b]	54.1
Economically active male population (%)	79.2[b]	80.2
Telephone lines (per 100 inhabitants)	8.6[c]	11.1[a]

Social indicators	1990/95
Contraceptive use (% of currently married women)	30

Environment	1990/91
Threatened species	49

a 1991. b 1980. c 1987.

Kenya

Region	Eastern Africa
Location (longitude, latitude)	1°28'S 36°82'E
Currency	Kenyan shilling
Population (1995 est., in thousands)	27885
Surface area (square kms)	580367
Population density (pop. per square km)	45
Sex ratio (females per 100 males)	100
Largest city (pop. in thousands)	Nairobi (1518)
UN membership date	16-Dec-1963
Major language(s)	English, Swahili, Kuyu, Luo, Kamba, Luhya, Nandi-kipsigis, Kisii

Economic indicators	1985	1994
Exchange rate (US$)	16.28	48.01
Consumer price index (1980=100)	187	305[a]
Balance of payments, current account (million US$)	−113	−98[b]
Tourist arrivals (in thousands)	695[c]	699[b]
GDP (million US$)	6138	8261[d]
GDP (per capita US$)	309	339[d]
Long-term rate of change in GDP (% per annum)	4.3	1.7[d]
Gross fixed capital formation (% of GDP)	17.4	18.8[d]
Economically active female population (%)	62.6[e]	55.4
Economically active male population (%)	90.7[e]	89.1
Annual growth of econ. active female pop. (%)	3.3[f]	...
Annual growth of econ. active male pop. (%)	3.9[f]	...
Labour force in industry (%)	21.6[e]	20.0[d]
Labour force in agriculture (%)	23.0[e]	18.9[d]
Agricultural production index (1979-1981=100)	146[c]	135[b]
Food production index (1979-1981=100)	146[c]	138[b]
Commercial energy production (1,000 Mt coal equiv)	130[e]	706[d]
Motor vehicles (per 1,000 population)	12	13[d]
Telephone lines (per 100 inhabitants)	0.7[g]	0.8[d]

Social indicators	1990/95
Growth rate of population (% per annum)	3.3
Age group 0-14 years (%)	47.4
Age group 60+ years (women and men, %)	2.4/2.0
Life expectancy at birth (women and men, years)	61/57
Infant mortality rate (per 1,000 births)	66
Total fertility rate (births per woman)	6
Contraceptive use (% of currently married women)	33
Urban population (%)	28
Urban population growth rate (% per annum)	6.6
Rural population growth rate (% per annum)	2.2
Foreign-born (1985,%)	0.8
Refugees	401900
Government education expenditure (% of GDP)	6
Primary-secondary gross enrolment ratio (f and m, per 100)	74/79
Third-level students (per 100,000 population)	135[h]
Newspaper circulation (per 1,000 population)	15
Television receivers (per 1,000 population)	10
Parliamentary seats (women and men, %)	3/97

Environment	1990/91
Threatened species	47
Forested area (%)	4.0
CO_2 emissions (10,000 Mt)	1323
Energy consumption (1,000 Mt coal equiv)	109
Precipitation (mm)	926[g]
Average temperature (January and July, centigrade)	17.8/14.9[e]

a April 1994. b 1992. c 1988. d 1991. e 1980. f 1990. g 1987. h 1989.

Kiribati

Region	Oceania-Micronesia
Location (longitude, latitude)	1°25'N 173°20'E[a]
Currency	Australian dollar
Population (1995 est., in thousands)	56
Surface area (square kms)	726
Population density (pop. per square km)	91
Sex ratio (females per 100 males)	103
Largest city (pop. in thousands)	Tarawa (25)[a]
Major language(s)	English, Kiribati

Economic indicators	1985	1994
Exchange rate (US$)	1.47	1.35
Consumer price index (1980=100)	131	181
Balance of payments, current account (million US$)	6	12[b]
Tourist arrivals (in thousands)	3[c]	4[b]
GDP (million US$)	20	39[d]
GDP (per capita US$)	306	544[d]
Long-term rate of change in GDP (% per annum)	-9.3	3.9[d]
Telephone lines (per 100 inhabitants)	1.4[f]	1.8[d]

Largest export industries		Major trading partners			1992
	(% of exports)		(% of exports)		(% of imports)
Agriculture	83	Bangladesh	67	Australia	38
Other manufacturing	16	United States	12	Japan	23
Basic metal industry	2	Australia	5	Fiji	11

Social indicators	1990/95
Age group 0-14 years (%)	41.1[e]
Age group 60+ years (women and men, %)	3.3/2.5[g]
Contraceptive use (% of currently married women)	37[ch]
Urban population (%)	39
Urban population growth rate (% per annum)	3.8
Rural population growth rate (% per annum)	1.2
Foreign-born (1985, %)	4.0
Intentional homicides (1986, per 100,000 population)	32
Parliamentary seats (women and men, %)	0/100

Environment	1990/91
Threatened species	7
Forested area (%)	2.8
CO_2 emissions (10,000 Mt)	6
Energy consumption (1,000 Mt coal equiv)	139
Average temperature (January and July, centigrade)	27.7/27.8[e]

a Capital city. b 1992. c 1988. d 1991. e 1980. f 1987. g 1978. h Source: NatSTB.

Korea, Democratic People's Republic of

Region Eastern Asia
Location (longitude, latitude) 39°01'N 125°74'E
Currency won
Population (1995 est., in thousands) 23922
Sex ratio (females per 100 males) 103
Largest city (pop. in thousands) Pyongyang (2230)
UN membership date 17-Sep-1991
Major language(s) Korean

Economic indicators	1985	1994
GDP (million US$)	16700	21310[a]
GDP (per capita US$)	840	960[a]
Long-term rate of change in GDP (% per annum)	9.6	0.0[a]
Economically active female population (%)	63.2[b]	66.2
Economically active male population (%)	81.2[b]	83.3
Annual growth of econ. active female pop. (%)	3.1[c]	...
Annual growth of econ. active male pop. (%)	2.9[c]	...
Agricultural production index (1979-1981=100)	126[d]	115[e]
Food production index (1979-1981=100)	125[d]	114[e]
Commercial energy production (1,000 Mt coal equiv)	44764[b]	84100[a]
Telephone lines (per 100 inhabitants)	3.3[f]	3.6[a]

Social indicators	1990/95
Growth rate of population (% per annum)	1.9
Age group 0-14 years (%)	29.1
Age group 60+ years (women and men, %)	4.5/2.5
Life expectancy at birth (women and men, years)	74/68
Infant mortality rate (per 1,000 births)	24
Total fertility rate (births per woman)	2
Urban population (%)	61
Urban population growth rate (% per annum)	2.4
Rural population growth rate (% per annum)	1.2
Foreign-born (1985,%)	0.2
Newspaper circulation (per 1,000 population)	230
Television receivers (per 1,000 population)	15
Parliamentary seats (women and men, %)	20/80

Environment	1990/91
Threatened species	30
Forested area (%)	74.4
CO_2 emissions (10,000 Mt)	66385
Energy consumption (1,000 Mt coal equiv)	4196

a 1991. b 1980. c 1990. d 1988. e 1992. f 1987.

Korea, Republic of

Region	Eastern Asia
Location (longitude, latitude)	37°58'N 126°98'E
Currency	won
Population (1995 est., in thousands)	45182
Surface area (square kms)	99016
Population density (pop. per square km)	437
Sex ratio (females per 100 males)	98
Largest city (pop. in thousands)	Seoul (10979)
UN membership date	17-Sep-1991
Major language(s)	Korean

Economic indicators	1985	1994
Exchange rate (US$)	890.20	798.90
Consumer price index (1980=100)	141	240
Industrial production index (1980 = 100)	165	392a
Unemployment (%)	...	2.2b
Balance of payments, current account (million US$)	−887	−4529c
Tourist arrivals (in thousands)	2340d	3231c
GDP (million US$)	92925	282971c
GDP (per capita US$)	2277	6462c
Long-term rate of change in GDP (% per annum)	6.9	8.4e
Gross fixed capital formation (% of GDP)	28.2	38.0e
Economically active female population (%)	39.5f	40.9
Economically active male population (%)	76.0f	78.8
Annual growth of econ. active female pop. (%)	2.9g	...
Annual growth of econ. active male pop. (%)	2.6g	...
Labour force in industry (%)	29.0f	34.6c
Labour force in agriculture (%)	34.0f	16.0c
Agricultural production index (1979-1981=100)	114d	113c
Food production index (1979-1981=100)	116d	114c
Commercial energy production (1,000 Mt coal equiv)	13774f	31218e
Motor vehicles (per 1,000 population)	27	97e
Telephone lines (per 100 inhabitants)	20.6h	33.3e

Largest export industries		Major trading partners			1992
	(% of exports)		(% of exports)		(% of imports)
Metal manufacture	48	United States	24	Japan	24
Textiles	26	Japan	15	United States	23
Chemicals	12	Hong Kong	8	Saudi Arabia	5

Social indicators	1990/95
Growth rate of population (% per annum)	0.8
Age group 0-14 years (%)	23.3
Age group 60+ years (women and men, %)	5.3/3.5
Life expectancy at birth (women and men, years)	74/68
Infant mortality rate (per 1,000 births)	21
Total fertility rate (births per woman)	2
Contraceptive use (% of currently married women)	79
Urban population (%)	78
Urban population growth rate (% per annum)	2.3
Rural population growth rate (% per annum)	−3.6
Foreign-born (1985,%)	2.1
Refugees	100
Government education expenditure (% of GDP)	3
Primary-secondary gross enrolment ratio (f and m, per 100)	97/97
Third-level students (per 100,000 population)	3953
Newspaper circulation (per 1,000 population)	277
Television receivers (per 1,000 population)	208
Intentional homicides (1986, per 100,000 population)	1
Parliamentary seats (women and men, %)	1/99

Environment	1990/91
Threatened species	29
Forested area (%)	65.4
CO_2 emissions (10,000 Mt)	72229
Energy consumption (1,000 Mt coal equiv)	2977
Precipitation (mm)	1093i
Average temperature (January and July, centigrade)	−4.0/23.9i

a July 1994. b Labour force sample surveys. c 1992. d 1988. e 1991.
f 1980. g 1990. h 1987. i Inchon.

Kuwait

Region	Western Asia
Location (longitude, latitude)	29°30'N 47°45'E
Currency	Kuwaiti dinar
Population (1995 est., in thousands)	1604
Surface area (square kms)	17818
Population density (pop. per square km)	118
Sex ratio (females per 100 males)	97
Largest city (pop. in thousands)	Kuwait City (1090)
UN membership date	14-May-1963
Major language(s)	Arabic

Economic indicators	1985	1994
Exchange rate (US$)	0.29	0.30
Consumer price index (1980=100)	124	135[a]
Balance of payments, current account (million US$)	5150	−873[b]
Tourist arrivals (in thousands)	80[c]	65[b]
GDP (million US$)	21429	24235[d]
GDP (per capita US$)	12458	11618[d]
Long-term rate of change in GDP (% per annum)	−4.3	−44.0[d]
Gross fixed capital formation (% of GDP)	19.8	...
Economically active female population (%)	20.3[e]	27.2
Economically active male population (%)	85.8[e]	82.9
Annual growth of econ. active female pop. (%)	10.0[a]	...
Annual growth of econ. active male pop. (%)	5.9[a]	...
Commercial energy production (1,000 Mt coal equiv)[f]	134712[e]	14858[d]
Motor vehicles (per 1,000 population)	324	...
Telephone lines (per 100 inhabitants)	13.1[g]	16.1[d]

Social indicators	1990/95
Growth rate of population (% per annum)	−5.8
Age group 0-14 years (%)	41.1
Age group 60+ years (women and men, %)	1.2/1.8
Life expectancy at birth (women and men, years)	78/73
Infant mortality rate (per 1,000 births)	14
Total fertility rate (births per woman)	4
Contraceptive use (% of currently married women)	35[g]
Urban population (%)	97
Urban population growth rate (% per annum)	−5.6
Rural population growth rate (% per annum)	−12.3
Foreign-born (1985,%)	59.4
Refugees	124900
Government education expenditure (% of GDP)	5[h]
Primary-secondary gross enrolment ratio (f and m, per 100)	86/89[i]
Third-level students (per 100,000 population)	1384[c]
Newspaper circulation (per 1,000 population)	210
Television receivers (per 1,000 population)	283
Intentional homicides (1986, per 100,000 population)	1
Parliamentary seats (women and men, %)	0/100

Environment	1990/91
Threatened species	10
Forested area (%)	0.1
CO_2 emissions (10,000 Mt)	3232[f]
Energy consumption (1,000 Mt coal equiv)[f]	1925
Precipitation (mm)	111[j]
Average temperature (January and July, centigrade)	13.9/36.7[e]

a 1990. b 1992. c 1988. d 1991. e 1980. f Includes part of the Neutral Zone. g 1987. h 1986. i 1989. j Shuwaikh.

Kyrgyzstan

Region Central Asia
Location (longitude, latitude) 42°53'N 76°46'E[a]
Currency rouble
Surface area (square kms) 198500
Largest city (pop. in thousands) Bishkek (629)[a]
UN membership date 02-Mar-1992
Major language(s) Kirghiz, Russian, Uzbec

Economic indicators	1985	1994
Exchange rate (US$)	...	8.80
GDP (million US$)	7301	8945[b]
GDP (per capita US$)	1819	2007[b]
Long-term rate of change in GDP (% per annum)	−5.3	−10.4[b]
Economically active female population (%)	55.6[c]	57.7
Economically active male population (%)	76.1[c]	77.9
Telephone lines (per 100 inhabitants)	5.8[d]	7.3[b]

Social indicators	1990/95
Contraceptive use (% of currently married women)	31

Environment	1990/91
Threatened species	17

a Capital city. b 1991. c 1980. d 1987.

Lao People's Democratic Republic

Region South-eastern Asia
Location (longitude, latitude) 17°57'N 102°34'E[a]
Currency new kip
Population (1995 est., in thousands) 4882
Surface area (square kms) 236800
Population density (pop. per square km) 18
Sex ratio (females per 100 males) 103
Largest city (pop. in thousands) Vientiane (415)[a]
UN membership date 14-Dec-1955
Major language(s) Lao, Thai, Kmhu

Economic indicators	1985	1994
Exchange rate (US$)	95.00	718.00
Balance of payments, current account (million US$)	−94	−41[b]
Tourist arrivals (in thousands)	25[c]	25[b]
GDP (million US$)	2496	1042[d]
GDP (per capita US$)	695	240[d]
Long-term rate of change in GDP (% per annum)	5.1	5.0[d]
Economically active female population (%)	78.0[e]	68.2
Economically active male population (%)	89.9[e]	88.4
Annual growth of econ. active female pop. (%)	1.6[f]	...
Annual growth of econ. active male pop. (%)	2.0[f]	...
Agricultural production index (1979-1981=100)	158[c]	160[b]
Food production index (1979-1981=100)	158[c]	160[b]
Commercial energy production (1,000 Mt coal equiv)	114[e]	115[d]
Telephone lines (per 100 inhabitants)	0.2[g]	0.1[d]

Social indicators	1990/95
Growth rate of population (% per annum)	3.0
Age group 0-14 years (%)	44.8
Age group 60+ years (women and men, %)	2.6/2.2
Life expectancy at birth (women and men, years)	53/50
Infant mortality rate (per 1,000 births)	97
Total fertility rate (births per woman)	7
Urban population (%)	22
Urban population growth rate (% per annum)	6.1
Rural population growth rate (% per annum)	2.2
Foreign-born (1985, %)	0.4
Government education expenditure (% of GDP)	1[c]
Primary-secondary gross enrolment ratio (f and m, per 100)	51/70
Third-level students (per 100,000 population)	118[h]
Newspaper circulation (per 1,000 population)	3
Television receivers (per 1,000 population)	6
Parliamentary seats (women and men, %)	9/91

Environment	1990/91
Threatened species	65
Forested area (%)	53.6
CO_2 emissions (10,000 Mt)	69
Energy consumption (1,000 Mt coal equiv)	38
Precipitation (mm)	1683[g]
Average temperature (January and July, centigrade)	21.5/27.7[e]

a Capital city. b 1992. c 1988. d 1991. e 1980. f 1990. g 1987. h 1989.

Latvia

Region Northern Europe
Location (longitude, latitude) 56°93'N 24°10'E
Currency rouble
Surface area (square kms) 64500
Largest city (pop. in thousands) Riga (921)
UN membership date 17-Sep-1991
Major language(s) Latvian, Russian

Economic indicators	1985	1994
Exchange rate (US$)	...	0.60
Consumer price index (1980=100)	...	5556[a]
GDP (million US$)	9934	16436[b]
GDP (per capita US$)	3802	6119[b]
Long-term rate of change in GDP (% per annum)	−0.2	−8.3[b]
Gross fixed capital formation (% of GDP)	6.2[b]	...
Economically active female population (%)	58.9[c]	57.7
Economically active male population (%)	78.5[c]	79.5
Motor vehicles (per 1,000 population)	110	153[b]
Telephone lines (per 100 inhabitants)	21.5[d]	23.9[b]

Social indicators	1990/95
Urban population (%)	73
Urban population growth rate (% per annum)	0.2
Rural population growth rate (% per annum)	−1.4
Television receivers (per 1,000 population)	372
Intentional homicides (1986, per 100,000 population)	9[e]

Environment	1990/91
Threatened species	33

a February 1994. b 1991. c 1980. d 1987. e Source: DYB92.

101

Lebanon

Region	Western Asia
Location (longitude, latitude)	33°90'N 35°50'E
Currency	Lebanese pound
Population (1995 est., in thousands)	3028
Surface area (square kms)	10400
Population density (pop. per square km)	264
Sex ratio (females per 100 males)	105
Largest city (pop. in thousands)	Beirut (1563)
UN membership date	24-Oct-1945
Major language(s)	Arabic

Economic indicators	1985	1994
Exchange rate (US$)	18.10	1666.00
GDP (million US$)	2071	1706[a]
GDP (per capita US$)	776	613[a]
Long-term rate of change in GDP (% per annum)	-16.8	3.0[a]
Economically active female population (%)	19.7[b]	25.0
Economically active male population (%)	73.0[b]	73.9
Annual growth of econ. active female pop. (%)	3.8[c]	...
Annual growth of econ. active male pop. (%)	1.2[c]	...
Agricultural production index (1979-1981=100)	138[d]	196[e]
Food production index (1979-1981=100)	141[d]	202[e]
Commercial energy production (1,000 Mt coal equiv)	104[b]	69[a]
Telephone lines (per 100 inhabitants)	13.7[c]	11.1[a]

Social indicators	1990/95
Growth rate of population (% per annum)	2.0
Age group 0-14 years (%)	34.2
Age group 60+ years (women and men, %)	4.5/3.8
Life expectancy at birth (women and men, years)	71/67
Infant mortality rate (per 1,000 births)	34
Total fertility rate (births per woman)	3
Urban population (%)	87
Urban population growth rate (% per annum)	2.8
Rural population growth rate (% per annum)	-2.7
Foreign-born (1985, %)	10.4
Refugees	6000
Primary-secondary gross enrolment ratio (f and m, per 100)	83/84[d]
Third-level students (per 100,000 population)	3071
Newspaper circulation (per 1,000 population)	117
Television receivers (per 1,000 population)	325
Parliamentary seats (women and men, %)	2/98

Environment	1990/91
Threatened species	13
Forested area (%)	7.7
CO_2 emissions (10,000 Mt)	2282
Energy consumption (1,000 Mt coal equiv)	1370
Precipitation (mm)	517[f]
Average temperature (January and July, centigrade)	13.9/26.2[b]

a 1991. b 1980. c 1990. d 1988. e 1992. f 1987.

Lesotho

Region	Western Africa
Location (longitude, latitude)	29°28'S 27°30'E[a]
Currency	maloti
Population (1995 est., in thousands)	1977
Surface area (square kms)	30355
Population density (pop. per square km)	60
Sex ratio (females per 100 males)	105
Largest city (pop. in thousands)	Maseru (170)[a]
UN membership date	17-Oct-1966
Major language(s)	English, Sotho, Zulu

Economic indicators	1985	1994
Exchange rate (US$)	2.56	3.56
Consumer price index (1980=100)	186	484
Balance of payments, current account (million US$)	−12	38[b]
Tourist arrivals (in thousands)	110[c]	155[b]
GDP (million US$)	252	596[d]
GDP (per capita US$)	163	333[d]
Long-term rate of change in GDP (% per annum)	3.5	0.0[d]
Gross fixed capital formation (% of GDP)	49.6	77.2[e]
Economically active female population (%)	70.9[f]	62.5
Economically active male population (%)	91.8[f]	90.4
Annual growth of econ. active female pop. (%)	1.5[e]	...
Annual growth of econ. active male pop. (%)	2.5[e]	...
Agricultural production index (1979-1981=100)	114[c]	81[b]
Food production index (1979-1981=100)	116[c]	79[b]
Motor vehicles (per 1,000 pop.)	13	
Telephone lines (per 100 inhabitants)	0.7[g]	0.6[d]

Social indicators	1990/95
Growth rate of population (% per annum)	2.5
Age group 0-14 years (%)	40.7
Age group 60+ years (women and men, %)	3.4/2.6
Life expectancy at birth (women and men, years)	63/58
Infant mortality rate (per 1,000 births)	79
Total fertility rate (births per woman)	5
Contraceptive use (% of currently married women)	23
Urban population (%)	23
Urban population growth rate (% per annum)	6.0
Rural population growth rate (% per annum)	1.5
Foreign-born (1985,%)	1.3
Refugees	100
Government education expenditure (% of GDP)	5
Primary-secondary gross enrolment ratio (f and m, per 100)	84/70
Third-level students (per 100,000 population)	333[c]
Newspaper circulation (per 1,000 population)	11
Television receivers (per 1,000 population)	6
Parliamentary seats (women and men, %)	2/98

Environment	1990/91
Threatened species	12

a Capital city. b 1992. c 1988. d 1991. e 1990. f 1980. g 1987.

Liberia

Region	Western Africa
Location (longitude, latitude)	6°20'N 10°46'W[a]
Currency	Liberian dollar
Population (1995 est., in thousands)	3039
Surface area (square kms)	111369
Population density (pop. per square km)	24
Sex ratio (females per 100 males)	98
Largest city (pop. in thousands)	Monrovia (668)[a]
UN membership date	02-Nov-1945
Major language(s)	English, Kpelle, Bassa, Grebo, Dan, Mano, Loma

Economic indicators	1985	1994
Exchange rate (US$)	1.00[b]	1.00[b]
Consumer price index (1980=100)	119	162[c]
Balance of payments, current account (million US$)	56	...
GDP (million US$)	1055	1037[d]
GDP (per capita US$)	480	390[d]
Long-term rate of change in GDP (% per annum)	−2.0	−10.0[d]
Gross fixed capital formation (% of GDP)	12.0	...
Economically active female population (%)	40.6[e]	35.5
Economically active male population (%)	89.4[e]	87.2
Annual growth of econ. active female pop. (%)	2.2[c]	...
Annual growth of econ. active male pop. (%)	2.6[c]	...
Agricultural production index (1979-1981=100)	119[f]	76[g]
Food production index (1979-1981=100)	122[f]	91[g]
Commercial energy production (1,000 Mt coal equiv)	41[e]	20[d]
Motor vehicles (per 1,000 population)	14	...
Telephone lines (per 100 inhabitants)	0.3[h]	0.1[d]

Social indicators	1990/95
Growth rate of population (% per annum)	3.3
Age group 0-14 years (%)	46.0
Age group 60+ years (women and men, %)	2.8/2.6
Life expectancy at birth (women and men, years)	57/54
Infant mortality rate (per 1,000 births)	126
Total fertility rate (births per woman)	7
Contraceptive use (% of currently married women)	6[i]
Urban population (%)	51
Urban population growth rate (% per annum)	5.5
Rural population growth rate (% per annum)	1.3
Foreign-born (1985,%)	4.6
Refugees	100000
Government education expenditure (% of GDP)	5[j]
Primary-secondary gross enrolment ratio (f and m, per 100)	25/48[e]
Third-level students (per 100,000 population)	220
Newspaper circulation (per 1,000 population)	14
Television receivers (per 1,000 population)	18
Intentional homicides (1986, per 100,000 population)	2
Parliamentary seats (women and men, %)	6/94

Environment	1990/91
Threatened species	34
Forested area (%)	17.8
CO_2 emissions (10,000 Mt)	75
Energy consumption (1,000 Mt coal equiv)	57

a Capital city. b Fixed rate. c 1990. d 1991. e 1980. f 1988. g 1992. h 1987. i 1986. j 1985.

Libyan Arab Jamahiriya

Region	Northern Africa
Location (longitude, latitude)	32°88'N 13°19'E
Currency	Libyan dinar
Population (1995 est., in thousands)	5407
Surface area (square kms)	1759540
Population density (pop. per square km)	3
Sex ratio (females per 100 males)	92
Largest city (pop. in thousands)	Tripoli (2595)
UN membership date	14-Dec-1955
Major language(s)	Arabic

Economic indicators	1985	1994
Exchange rate (US$)	0.30	0.30
Balance of payments, current account (million US$)	1906	2201[a]
Tourist arrivals (in thousands)	98[b]	89[c]
GDP (million US$)	27963	44967[d]
GDP (per capita US$)	7386	9551[d]
Long-term rate of change in GDP (% per annum)	8.3	5.0[d]
Economically active female population (%)	7.2[e]	9.4
Economically active male population (%)	80.2[e]	76.8
Annual growth of econ. active female pop. (%)	5.9[a]	...
Annual growth of econ. active male pop. (%)	3.6[a]	...
Agricultural production index (1979-1981=100)	106[b]	186[c]
Food production index (1979-1981=100)	106[b]	187[c]
Commercial energy production (1,000 Mt coal equiv)	132809[e]	116564[d]
Motor vehicles (per 1,000 population)	166	169[a]
Telephone lines (per 100 inhabitants)	4.2[f]	5.7[d]

Social indicators	1990/95
Growth rate of population (% per annum)	3.5
Age group 0-14 years (%)	45.4
Age group 60+ years (women and men, %)	2.0/2.4
Life expectancy at birth (women and men, years)	65/62
Infant mortality rate (per 1,000 births)	68
Total fertility rate (births per woman)	6
Urban population (%)	86
Urban population growth rate (% per annum)	4.3
Rural population growth rate (% per annum)	−1.1
Foreign-born (1985,%)	14.3
Government education expenditure (% of GDP)	8[g]
Third-level students (per 100,000 population)	1548
Newspaper circulation (per 1,000 population)	15
Television receivers (per 1,000 population)	99

Environment	1990/91
Threatened species	17
Forested area (%)	0.4
CO_2 emissions (10,000 Mt)	11738
Energy consumption (1,000 Mt coal equiv)	4328
Precipitation (mm)	253[h]
Average temperature (January and July, centigrade)	10.9/27.1[e]

a 1990. b 1988. c 1992. d 1991. e 1980. f 1987. g 1986. h Idris.

Liechtenstein

Region	Western Europe
Location (longitude, latitude)	47°14'N 9°52'E[a]
Currency	Swiss franc
Surface area (square kms)	160
Population density (pop. per square km)	175
Largest city (pop. in thousands)	Vaduz (6)[a]
UN membership date	18-Sep-1990
Major language(s)	German

Economic indicators	1985	1994
Tourist arrivals (in thousands)	72[b]	72[c]
GDP (million US$)	529	1504[d]
GDP (per capita US$)	19596	53724[d]
Long-term rate of change in GDP (% per annum)	4.2	0.0[d]
Telephone lines (per 100 inhabitants)	51.2[e]	62.5[d]

Social indicators	1990/95
Urban population (%)	21
Urban population growth rate (% per annum)	1.3
Rural population growth rate (% per annum)	−0.1
Foreign-born (1985,%)	35.1
Newspaper circulation (per 1,000 population)	307
Television receivers (per 1,000 population)	345

Environment	1990/91
Threatened species	10
Forested area (%)	18.8

a Capital city. b 1988. c 1992. d 1991. e 1987.

Lithuania

Region	Northern Europe
Location (longitude, latitude)	54°41'N 25°19'E[a]
Currency	rouble
Surface area (square kms)	65200
Population density (pop. per square km)	57
Largest city (pop. in thousands)	Vilnius (598)[a]
UN membership date	17-Sep-1991
Major language(s)	Lithuanian, Russian, Polish

Economic indicators	1985	1994
Exchange rate (US$)	...	4.00
Consumer price index (1980=100)	106	3546
GDP (million US$)	13569	21802[b]
GDP (per capita US$)	3783	5820[b]
Long-term rate of change in GDP (% per annum)	9.6	−13.5[b]
Economically active female population (%)	57.6[c]	56.3
Economically active male population (%)	77.9[c]	79.6
Motor vehicles (per 1,000 population)	123	172[b]
Telephone lines (per 100 inhabitants)	17.4[d]	21.6[b]

Largest export industries	Major trading partners			1992
(% of exports)	(% of exports)		(% of imports)	
...	Russian Fed.	28	Russian Fed.	58
...	Ukraine	16	Ukraine	8
...	Belarus	12	Belarus	6

Social indicators	1990/95
Urban population (%)	72
Urban population growth rate (% per annum)	1.2
Rural population growth rate (% per annum)	−2.0
Television receivers (per 1,000 population)	374
Intentional homicides (1986, per 100,000 population)	8[e]

Environment	1990/91
Threatened species	33

a Capital city. b 1991. c 1980. d 1987. e Source: DYB92.

Luxembourg

Region Western Europe
Location (longitude, latitude) 49°61'N 6°12'E[a]
Currency Luxembourg franc
Population (1995 est., in thousands) 386
Surface area (square kms) 2586
Population density (pop. per square km) 144
Sex ratio (females per 100 males) 103
Largest city (pop. in thousands) Luxembourg-ville (83)[a]
UN membership date 24-Oct-1945
Major language(s) French, Luxemburgish

Economic indicators	1985	1994
Exchange rate (US$)	50.36	31.83
Consumer price index (1980=100)	140	113
Industrial production index (1980 = 100)	121	140[b]
Tourist arrivals (in thousands)	760[c]	796[d]
GDP (million US$)	3457	9336[e]
GDP (per capita US$)	9419	24896[e]
Long-term rate of change in GDP (% per annum)	2.9	3.1[e]
Gross fixed capital formation (% of GDP)	17.7	29.0[e]
Economically active female population (%)	32.5[f]	32.2
Economically active male population (%)	72.4[f]	72.8
Annual growth of econ. active female pop. (%)	1.8[g]	...
Annual growth of econ. active male pop. (%)	0.5[g]	...
Labour force in industry (%)	35.3[h]	30.4[g]
Labour force in agriculture (%)	4.7[h]	3.3[g]
Commercial energy production (1,000 Mt coal equiv)	12[f]	98[e]
Motor vehicles (per 1,000 population)	486	604[e]
Telephone lines (per 100 inhabitants)	43.8[i]	51.1[e]

Social indicators	1990/95
Growth rate of population (% per annum)	0.7
Age group 0-14 years (%)	17.6
Age group 60+ years (women and men, %)	11.7/7.8
Life expectancy at birth (women and men, years)	79/72
Infant mortality rate (per 1,000 births)	8
Total fertility rate (births per woman)	2
Urban population (%)	86
Urban population growth rate (% per annum)	1.1
Rural population growth rate (% per annum)	−1.7
Foreign-born (1985, %)	27.1
Refugees	2200
Government education expenditure (% of GDP)	4[j]
Primary-secondary gross enrolment ratio (f and m, per 100)	82/80[j]
Third-level students (per 100,000 population)	245[k]
Newspaper circulation (per 1,000 population)	383
Television receivers (per 1,000 population)	267
Intentional homicides (1986, per 100,000 population)	2[l]
Parliamentary seats (women and men, %)	13/87

Environment	1990/91
Threatened species	14
CO_2 emissions (10,000 Mt)	2814
Energy consumption (1,000 Mt coal equiv)	13757
Precipitation (mm)	740[i]
Average temperature (January and July, centigrade)	0.3/17.4[f]

a Capital city. b July 1994. c 1988. d 1992. e 1991. f 1980. g 1990.
h 1983. i 1987. j 1989. k 1985. l Source: DYB92.

Macau

Region	Eastern Asia
Location (longitude, latitude)	22°12'N 113°33'E[a]
Currency	pataca
Population (1995 est., in thousands)	444
Surface area (square kms)	16
Population density (pop. per square km)	31063
Sex ratio (females per 100 males)	94
Largest city (pop. in thousands)	Macau (457)[a]
Major language(s)	Chinese, Portuguese

Economic indicators	1985	1994
Consumer price index (1980=100)	102	182[b]
Tourist arrivals (in thousands)	2273[c]	3180[d]
Agricultural production index (1979-1981=100)	112[c]	73[d]
Food production index (1979-1981=100)	112[c]	73[d]
Motor vehicles (per 1,000 population)	70[g]	74[f]
Telephone lines (per 100 inhabitants)	13.2[h]	22.3[f]

Largest export industries	Major trading partners			1992	
(% of exports)	(% of exports)		(% of imports)		
Textiles 80	United States	35	Hong Kong	33	
Metal manufacture 8	Hong Kong	13	China	20	
Other manufacturing 6	Germany	12	Japan	18	

Social indicators	1990/95
Age group 0-14 years (%)	21.8
Age group 60+ years (women and men, %)	4.9/3.6[ci]
Life expectancy at birth (women and men, years)	80/75[j]
Infant mortality rate (per 1,000 births)	9
Urban population (%)	99
Urban population growth rate (% per annum)	2.9
Rural population growth rate (% per annum)	0.5
Foreign-born (1985,%)	39.0
Newspaper circulation (per 1,000 population)	518
Television receivers (per 1,000 population)	67
Intentional homicides (1986, per 100,000 population)	2[k]

Environment	1990/91
CO_2 emissions (10,000 Mt)	297
Energy consumption (1,000 Mt coal equiv)	1092
Precipitation (mm)	1846[h]
Average temperature (January and July, centigrade)	15.1/28.5[e]

a Capital city. b May 1994. c 1988. d 1992. e 1980. f 1991. g 1990.
h 1987. i Source DYB90. j 1985. k Source: DYB92.

Madagascar

Region	Eastern Africa
Location (longitude, latitude)	15°91'S 47°53'E[a]
Currency	Malagasy franc
Population (1995 est., in thousands)	14155
Surface area (square kms)	587041
Population density (pop. per square km)	20
Sex ratio (females per 100 males)	102
Largest city (pop. in thousands)	Antananarivo (687)[a]
UN membership date	20-Sep-1960
Major language(s)	Malagasy

Economic indicators	1985	1994
Exchange rate (US$)	635.79	3718.60
Consumer price index (1980=100)	249	1063
Balance of payments, current account (million US$)	−184	−136[b]
Tourist arrivals (in thousands)	35[c]	54[b]
GDP (million US$)	2345	2673[d]
GDP (per capita US$)	229	215[d]
Long-term rate of change in GDP (% per annum)	2.3	−6.6[d]
Economically active female population (%)	60.8[e]	53.4
Economically active male population (%)	90.1[e]	88.5
Annual growth of econ. active female pop. (%)	1.9[f]	...
Annual growth of econ. active male pop. (%)	2.3[f]	...
Agricultural production index (1979-1981=100)	116[c]	124[b]
Food production index (1979-1981=100)	117[c]	126[b]
Commercial energy production (1,000 Mt coal equiv)	18[e]	39[d]
Motor vehicles (per 1,000 population)	9	...
Telephone lines (per 100 inhabitants)	0.2[g]	0.3[d]

Social indicators	1990/95
Growth rate of population (% per annum)	3.3
Age group 0-14 years (%)	45.7
Age group 60+ years (women and men, %)	2.5/2.1
Life expectancy at birth (women and men, years)	57/54
Infant mortality rate (per 1,000 births)	110
Total fertility rate (births per woman)	7
Contraceptive use (% of currently married women)	17
Urban population (%)	27
Urban population growth rate (% per annum)	5.6
Rural population growth rate (% per annum)	2.4
Foreign-born (1985,%)	0.3
Government education expenditure (% of GDP)	2
Primary-secondary gross enrolment ratio (f and m, per 100)	52/55[h]
Third-level students (per 100,000 population)	298
Newspaper circulation (per 1,000 population)	4
Television receivers (per 1,000 population)	20
Intentional homicides (1986, per 100,000 population)	7

Environment	1990/91
Threatened species	127
Forested area (%)	26.5
CO_2 emissions (10,000 Mt)	293
Energy consumption (1,000 Mt coal equiv)	43
Precipitation (mm)	1270[g]
Average temperature (January and July, centigrade)	19.3/13.0[e]

a Capital city. b 1992. c 1988. d 1991. e 1980. f 1990. g 1987. h 1989.

Malawi

Region	Eastern Africa
Location (longitude, latitude)	13°58'S 33°42'E[a]
Currency	Malawi kwacha
Population (1995 est., in thousands)	11304
Surface area (square kms)	118484
Population density (pop. per square km)	72
Sex ratio (females per 100 males)	102
Largest city (pop. in thousands)	Lilongwe (310)[a]
UN membership date	01-Sep-1964
Major language(s)	Nyanja, English, Makua, Yao

Economic indicators	1985	1994
Exchange rate (US$)	1.68	7.39
Consumer price index (1980=100)	185	201[b]
Industrial production index (1980 = 100)	100	132[c]
Balance of payments, current account (million US$)	−125	...
Tourist arrivals (in thousands)	99[d]	135[c]
GDP (million US$)	1122	2191[e]
GDP (per capita US$)	153	219[e]
Long-term rate of change in GDP (% per annum)	7.8	0.0
Gross fixed capital formation (% of GDP)	13.5	12.0[f]
Economically active female population (%)	62.4[g]	54.7
Economically active male population (%)	89.9[g]	88.2
Annual growth of econ. active female pop. (%)	2.1[f]	...
Annual growth of econ. active male pop. (%)	2.9[f]	...
Labour force in industry (%)	20.9[g]	21.5[f]
Labour force in agriculture (%)	49.2[g]	40.0[f]
Agricultural production index (1979-1981=100)	114[d]	98[c]
Food production index (1979-1981=100)	110[d]	86[c]
Commercial energy production (1,000 Mt coal equiv)	48[g]	90[e]
Motor vehicles (per 1,000 population)	4	5[f]
Telephone lines (per 100 inhabitants)	0.3[h]	0.3[e]

Social indicators	1990/95
Growth rate of population (% per annum)	3.3
Age group 0-14 years (%)	49.2
Age group 60+ years (women and men, %)	2.3/1.8
Life expectancy at birth (women and men, years)	45/44
Infant mortality rate (per 1,000 births)	142
Total fertility rate (births per woman)	8
Contraceptive use (% of currently married women)	13[d]
Urban population (%)	14
Urban population growth rate (% per annum)	6.0
Rural population growth rate (% per annum)	2.9
Foreign-born (1985, %)	3.9
Refugees	1058500
Government education expenditure (% of GDP)	2
Primary-secondary gross enrolment ratio (f and m, per 100)	47/58[i]
Third-level students (per 100,000 population)	61[d]
Newspaper circulation (per 1,000 population)	3
Intentional homicides (1986, per 100,000 population)	2
Parliamentary seats (women and men, %)	12/88

Environment	1990/91
Threatened species	16
Forested area (%)	30.6
CO_2 emissions (10,000 Mt)	172
Energy consumption (1,000 Mt coal equiv)	37
Precipitation (mm)	849[h]
Average temperature (January and July, centigrade)	21.2/15.0[g]

a Capital city. b June 1994. c 1992. d 1988. e 1991. f 1990. g 1980. h 1987. i 1989.

Malaysia

Region South-eastern Asia
Location (longitude, latitude) 3°14'N 101°71'E
Currency ringgit
Population (1995 est., in thousands) 20125
Surface area (square kms) 329749
Population density (pop. per square km) 56
Sex ratio (females per 100 males) 98
Largest city (pop. in thousands) Kuala Lumpur (1684)
UN membership date 17-Sep-1957
Major language(s) Malay, Chinese, Tamil

Economic indicators	1985	1994
Exchange rate (US$)	2.43	2.57
Consumer price index (1980=100)	125	117
Industrial production index (1980 = 100)	138	362[a]
Balance of payments, current account (million US$)	−613	−2103[b]
Tourist arrivals (in thousands)	3624[c]	6016[d]
GDP (million US$)	31200	47104[e]
GDP (per capita US$)	1990	2568[e]
Long-term rate of change in GDP (% per annum)	−1.1	8.7[e]
Gross fixed capital formation (% of GDP)	29.8	35.5[e]
Economically active female population (%)	43.0[f]	45.3
Economically active male population (%)	82.7[f]	82.7
Annual growth of econ. active female pop. (%)	4.2[g]	...
Annual growth of econ. active male pop. (%)	3.1[g]	...
Labour force in industry (%)	23.8[f]	27.5[g]
Labour force in agriculture (%)	37.2[f]	26.0[g]
Agricultural production index (1979-1981=100)	164[c]	205[d]
Food production index (1979-1981=100)	187[c]	255[d]
Commercial energy production (1,000 Mt coal equiv)	19474[f]	66531[e]
Motor vehicles (per 1,000 population)	108	135[e]
Telephone lines (per 100 inhabitants)	6.8[h]	9.9

Social indicators	1990/95
Growth rate of population (% per annum)	2.4
Age group 0-14 years (%)	37.9
Age group 60+ years (women and men, %)	3.2/2.7
Life expectancy at birth (women and men, years)	73/69
Infant mortality rate (per 1,000 births)	14
Total fertility rate (births per woman)	4
Contraceptive use (% of currently married women)	48
Urban population (%)	47
Urban population growth rate (% per annum)	4.2
Rural population growth rate (% per annum)	0.8
Foreign-born (1985,%)	4.7
Refugees	10300
Government education expenditure (% of GDP)	5
Primary-secondary gross enrolment ratio (f and m, per 100)	76/74
Third-level students (per 100,000 population)	671
Newspaper circulation (per 1,000 population)	140
Television receivers (per 1,000 population)	149
Intentional homicides (1986, per 100,000 population)	2
Parliamentary seats (women and men, %)	5/95

Environment	1990/91
Threatened species	110
Forested area (%)	58.6
CO_2 emissions (10,000 Mt)	16702
Energy consumption (1,000 Mt coal equiv)	1534
Precipitation (mm)	2499[h]
Average temperature (January and July, centigrade)	26.8/27.1[f]

a July 1994. b 1993. c 1988. d 1992. e 1991. f 1980. g 1990. h 1987.

Maldives

Region	Southern Asia
Location (longitude, latitude)	4°10'N 73°30'E[a]
Currency	rufiyaa
Population (1995 est., in thousands)	248
Surface area (square kms)	298
Population density (pop. per square km)	748
Sex ratio (females per 100 males)	92
Largest city (pop. in thousands)	Male (63)[a]
UN membership date	21-Sep-1965
Major language(s)	Maldivian

Economic indicators	1985	1994
Exchange rate (US$)	7.13	11.86
Balance of payments, current account (million US$)	−6	−33[b]
Tourist arrivals (in thousands)	156[c]	236[b]
GDP (million US$)	84	152[d]
GDP (per capita US$)	459	689[d]
Long-term rate of change in GDP (% per annum)	13.8	4.7[d]
Economically active female population (%)	27.1[e]	24.7
Economically active male population (%)	83.2[e]	82.4
Agricultural production index (1979-1981=100)	121[c]	128[b]
Food production index (1979-1981=100)	121[c]	128[b]
Telephone lines (per 100 inhabitants)	2.9[f]	3.5[d]

Social indicators	1990/95
Growth rate of population (% per annum)	3.0
Age group 0-14 years (%)	44.0
Age group 60+ years (women and men, %)	2.8/2.8
Life expectancy at birth (women and men, years)	62/65
Infant mortality rate (per 1,000 births)	55
Total fertility rate (births per woman)	6
Urban population (%)	33
Urban population growth rate (% per annum)	5.5
Rural population growth rate (% per annum)	1.9
Foreign-born (1985,%)	1.4
Newspaper circulation (per 1,000 population)	8
Television receivers (per 1,000 population)	25
Parliamentary seats (women and men, %)	4/96

Environment	1990/91
Threatened species	5
Forested area (%)	3.3
CO_2 emissions (10,000 Mt)	26
Energy consumption (1,000 Mt coal equiv)	205

a Capital city. b 1992. c 1988. d 1991. e 1980. f 1990.

Mali

Region	Western Africa
Location (longitude, latitude)	12°65'N 8°01'W[a]
Currency	CFA franc
Population (1995 est., in thousands)	10797
Surface area (square kms)	1240192
Population density (pop. per square km)	8
Sex ratio (females per 100 males)	103
Largest city (pop. in thousands)	Bamako (738)[a]
UN membership date	28-Sep-1960
Major language(s)	English, Bambara, Ful, Senufo, Soninke, Malinke, Songhai

Economic indicators	1985	1994
Exchange rate (US$)	378.05	528.15
Consumer price index (1980=100)	101[b]	119[c]
Balance of payments, current account (million US$)	−129	−91[d]
Tourist arrivals (in thousands)	36[e]	38[d]
GDP (million US$)	1158	2451[f]
GDP (per capita US$)	146	258[f]
Long-term rate of change in GDP (% per annum)	8.5	−0.2[f]
Gross fixed capital formation (% of GDP)	19.2	20.0[f]
Economically active female population (%)	16.7[g]	15.1
Economically active male population (%)	91.2[g]	90.3
Annual growth of econ. active female pop. (%)	1.8[b]	...
Annual growth of econ. active male pop. (%)	2.2[b]	...
Agricultural production index (1979-1981=100)	128[e]	137[d]
Food production index (1979-1981=100)	127[e]	132[d]
Commercial energy production (1,000 Mt coal equiv)	10[g]	23[f]
Motor vehicles (per 1,000 population)	4	...
Telephone lines (per 100 inhabitants)	0.1[h]	0.1[f]

Social indicators	1990/95
Growth rate of population (% per annum)	3.2
Age group 0-14 years (%)	47.4
Age group 60+ years (women and men, %)	2.3/1.8
Life expectancy at birth (women and men, years)	48/44
Infant mortality rate (per 1,000 births)	159
Total fertility rate (births per woman)	7
Contraceptive use (% of currently married women)	5[h]
Urban population (%)	27
Urban population growth rate (% per annum)	5.7
Rural population growth rate (% per annum)	2.3
Foreign-born (1985, %)	1.4
Refugees	13100
Government education expenditure (% of GDP)	3[h]
Primary-secondary gross enrolment ratio (f and m, per 100)	11/20[i]
Third-level students (per 100,000 population)	73
Newspaper circulation (per 1,000 population)	1
Television receivers (per 1,000 population)	1
Parliamentary seats (women and men, %)	2/98

Environment	1990/91
Threatened species	20
Forested area (%)	5.6
CO_2 emissions (10,000 Mt)	119
Energy consumption (1,000 Mt coal equiv)	24
Precipitation (mm)	1099[h]
Average temperature (January and July, centigrade)	25.5/26.9[g]

a Capital city. b 1990. c March 1994. d 1992. e 1988. f 1991. g 1980. h 1987. i 1989.

India

Region	Southern Asia
Location (longitude, latitude)	18°95'N 72°66'E
Currency	Indian rupee
Population (1995 est., in thousands)	931044
Surface area (square kms)	3287590
Population density (pop. per square km)	258
Sex ratio (females per 100 males)	94
Largest city (pop. in thousands)	Bombay (12223)
UN membership date	30-Oct-1945
Major language(s)	Hindi, English, Bengali, Telugu, Marathi, Tamil

Economic indicators	1985	1994
Exchange rate (US$)	12.16	31.37
Consumer price index (1980=100)	156	359[a]
Industrial production index (1980 = 100)	140	225[b]
Balance of payments, current account (million US$)	−4177	−7037[c]
Tourist arrivals (in thousands)	1591[d]	1868[e]
GDP (million US$)	212016	268006[f]
GDP (per capita US$)	277	311[f]
Long-term rate of change in GDP (% per annum)	5.5	1.3[f]
Gross fixed capital formation (% of GDP)	20.7	23.1[c]
Economically active female population (%)	32.2[g]	28.5
Economically active male population (%)	84.8[g]	84.0
Annual growth of econ. active female pop. (%)	1.2[c]	...
Annual growth of econ. active male pop. (%)	2.3[c]	...
Labour force in industry (%)	38.7[g]	36.3[h]
Labour force in agriculture (%)	5.8[g]	5.6[h]
Agricultural production index (1979-1981=100)	138[d]	155[e]
Food production index (1979-1981=100)	140[d]	157[e]
Commercial energy production (1,000 Mt coal equiv)	114066[g]	264714[f]
Motor vehicles (per 1,000 population)	5	8[f]
Telephone lines (per 100 inhabitants)	0.5[i]	0.7[f]

Social indicators	1990/95
Growth rate of population (% per annum)	1.9
Age group 0-14 years (%)	34.9
Age group 60+ years (women and men, %)	3.8/3.8
Life expectancy at birth (women and men, years)	61/60
Infant mortality rate (per 1,000 births)	88
Total fertility rate (births per woman)	4
Contraceptive use (% of currently married women)	43[d]
Urban population (%)	27
Urban population growth rate (% per annum)	2.9
Rural population growth rate (% per annum)	1.6
Foreign-born (1985,%)	1.2
Refugees	258400
Government education expenditure (% of GDP)	3[i]
Primary-secondary gross enrolment ratio (f and m, per 100)	55/79
Third-level students (per 100,000 population)	581[j]
Newspaper circulation (per 1,000 population)	26[g]
Television receivers (per 1,000 population)	35
Intentional homicides (1986, per 100,000 population)	3
Parliamentary seats (women and men, %)	7/93

Environment	1990/91
Threatened species	172
Forested area (%)	20.3
CO_2 emissions (10,000 Mt)	192017
Energy consumption (1,000 Mt coal equiv)	336
Precipitation (mm)	715[k]
Average temperature (January and July, centigrade)	14.3/31.2[g]

a August 1994. b April 1994. c 1990. d 1988. e 1992. f 1991. g 1980. h 1989. i 1987. j 1985. k New Delhi.

Indonesia

Region South-eastern Asia
Location (longitude, latitude) 6°18'S 106°89'E
Currency rupiah
Population (1995 est., in thousands) 201477
Surface area (square kms) 1904569
Population density (pop. per square km) 99
Sex ratio (females per 100 males) 101
Largest city (pop. in thousands) Jakarta (9206)
UN membership date 28-Sep-1950
Major language(s) Indonesian, Javanese, Sudanese

Economic indicators	1985	1994
Exchange rate (US$)	1125.00	2181.00
Consumer price index (1980=100)	159	156[a]
Balance of payments, current account (million US$)	−1923	−3679[b]
Tourist arrivals (in thousands)	1301[c]	3064[b]
GDP (million US$)	87339	116476[d]
GDP (per capita US$)	522	620[d]
Long-term rate of change in GDP (% per annum)	2.5	6.6
Gross fixed capital formation (% of GDP)	23.1	26.9[d]
Economically active female population (%)	36.9[e]	36.6
Economically active male population (%)	84.7[e]	83.0
Annual growth of econ. active female pop. (%)	2.6[f]	...
Annual growth of econ. active male pop. (%)	2.3[f]	...
Labour force in industry (%)	13.1[e]	14.1[b]
Labour force in agriculture (%)	56.3[e]	54.9[b]
Agricultural production index (1979-1981=100)	147[c]	177[b]
Food production index (1979-1981=100)	150[c]	182[b]
Commercial energy production (1,000 Mt coal equiv)	132256[e]	180506[d]
Motor vehicles (per 1,000 population)	12	17[d]
Telephone lines (per 100 inhabitants)	0.4[g]	0.7[d]

Largest export industries	Major trading partners		1992
(% of exports)	(% of exports)	(% of imports)	
Mining, quarrying 32	Japan 32	Japan	22
Textiles 22	United States 13	United States	14
Wood, wood products 12	Singapore 9	Germany	8

Social indicators	1990/95
Growth rate of population (% per annum)	1.8
Age group 0-14 years (%)	33.4
Age group 60+ years (women and men, %)	3.7/3.3
Life expectancy at birth (women and men, years)	65/61
Infant mortality rate (per 1,000 births)	65
Total fertility rate (births per woman)	3
Contraceptive use (% of currently married women)	50
Urban population (%)	33
Urban population growth rate (% per annum)	4.2
Rural population growth rate (% per annum)	0.7
Foreign-born (1985, %)	0.7
Refugees	15600
Government education expenditure (% of GDP)	1[c]
Primary-secondary gross enrolment ratio (f and m, per 100)	78/85[h]
Third-level students (per 100,000 population)	375[i]
Newspaper circulation (per 1,000 population)	28
Television receivers (per 1,000 population)	59
Intentional homicides (1986, per 100,000 population)	1
Parliamentary seats (women and men, %)	12/88

Environment	1990/91
Threatened species	354
Forested area (%)	59.6
CO_2 emissions (10,000 Mt)	46525[j]
Energy consumption (1,000 Mt coal equiv)	348
Precipitation (mm)	1755[g]
Average temperature (January and July, centigrade)	26.2/26.7[e]

a June 1994. b 1992. c 1988. d 1991. e 1980. f 1990. g 1987. h 1989.
i 1981. j Includes East Timor.

Iran, Islamic Republic of

Region	Southern Asia
Location (longitude, latitude)	35°69'N 51°41'E
Currency	Iranian rial
Population (1995 est., in thousands)	66720
Surface area (square kms)	1648000
Population density (pop. per square km)	34
Sex ratio (females per 100 males)	97
Largest city (pop. in thousands)	Teheran (6654)
UN membership date	24-Oct-1945
Major language(s)	Persian, Azerbaijani, Kurdish

Economic indicators	1985	1994
Exchange rate (US$)	84.23	67.04
Consumer price index (1980=100)	84	162[a]
Balance of payments, current account (million US$)	-476	-7909[b]
Tourist arrivals (in thousands)	67[c]	185[d]
GDP (million US$)	173253	649018[b]
GDP (per capita US$)	3542	10826[b]
Long-term rate of change in GDP (% per annum)	1.8	4.6[b]
Gross fixed capital formation (% of GDP)	17.5	15.5[e]
Economically active female population (%)	14.5[f]	18.6
Economically active male population (%)	81.0[f]	78.1
Annual growth of econ. active female pop. (%)	5.2[e]	...
Annual growth of econ. active male pop. (%)	3.2[e]	...
Agricultural production index (1979-1981=100)	144[c]	189[d]
Food production index (1979-1981=100)	144[c]	192[d]
Commercial energy production (1,000 Mt coal equiv)	116046[f]	284345[b]
Motor vehicles (per 1,000 population)	53	...
Telephone lines (per 100 inhabitants)	2.8[g]	4.1[b]

Social indicators	1990/95
Growth rate of population (% per annum)	2.7
Age group 0-14 years (%)	45.9
Age group 60+ years (women and men, %)	2.9/3.0
Life expectancy at birth (women and men, years)	68/67
Infant mortality rate (per 1,000 births)	40
Total fertility rate (births per woman)	6
Contraceptive use (% of currently married women)	65
Urban population (%)	60
Urban population growth rate (% per annum)	3.9
Rural population growth rate (% per annum)	1.0
Foreign-born (1985,%)	5.9
Refugees	4150700
Government education expenditure (% of GDP)	4
Primary-secondary gross enrolment ratio (f and m, per 100)	75/90
Third-level students (per 100,000 population)	1061
Newspaper circulation (per 1,000 population)	26
Television receivers (per 1,000 population)	63
Parliamentary seats (women and men, %)	3/97

Environment	1990/91
Threatened species	52
Forested area (%)	10.9
CO_2 emissions (10,000 Mt)	60688
Energy consumption (1,000 Mt coal equiv)	1654
Precipitation (mm)	208[g]
Average temperature (January and July, centigrade)	3.5/29.5[f]

a March 1994. b 1991. c 1988. d 1992. e 1990. f 1980. g 1987.

Iraq

Region	Western Asia
Location (longitude, latitude)	33°34'N 44°43'E
Currency	Iraqi dinar
Population (1995 est., in thousands)	21224
Surface area (square kms)	438317
Population density (pop. per square km)	45
Sex ratio (females per 100 males)	96
Largest city (pop. in thousands)	Baghdad (4044)
UN membership date	21-Dec-1945
Major language(s)	Arabic, Kurdish

Economic indicators	1985	1994
Exchange rate (US$)	0.31	0.31
Consumer price index (1980=100)	199	161[a]
Tourist arrivals (in thousands)	1209[b]	504[c]
GDP (million US$)	49819	62628[d]
GDP (per capita US$)	3252	3353[d]
Long-term rate of change in GDP (% per annum)	0.0	−46.0[d]
Gross fixed capital formation (% of GDP)	27.8	...
Economically active female population (%)	17.3[e]	22.8
Economically active male population (%)	78.9[e]	77.4
Annual growth of econ. active female pop. (%)	10.3[a]	...
Annual growth of econ. active male pop. (%)	3.3[a]	...
Agricultural production index (1979-1981=100)	127[b]	122[c]
Food production index (1979-1981=100)	129[b]	123[c]
Commercial energy production (1,000 Mt coal equiv)	188237[e]	20685[d]
Motor vehicles (per 1,000 population)	35	58[a]
Telephone lines (per 100 inhabitants)	4.1[f]	3.6[d]

Social indicators	1990/95
Growth rate of population (% per annum)	3.2
Age group 0-14 years (%)	43.8
Age group 60+ years (women and men, %)	2.4/2.2
Life expectancy at birth (women and men, years)	67/65
Infant mortality rate (per 1,000 births)	58
Total fertility rate (births per woman)	6
Contraceptive use (% of currently married women)	14[g]
Urban population (%)	75
Urban population growth rate (% per annum)	4.0
Rural population growth rate (% per annum)	1.1
Foreign-born (1985,%)	3.3
Refugees	95000
Primary-secondary gross enrolment ratio (f and m, per 100)	72/92
Third-level students (per 100,000 population)	1188[b]
Newspaper circulation (per 1,000 population)	36
Television receivers (per 1,000 population)	72
Parliamentary seats (women and men, %)	11/89

Environment	1990/91
Threatened species	30
Forested area (%)	4.3
CO_2 emissions (10,000 Mt)	11560
Energy consumption (1,000 Mt coal equiv)	970
Precipitation (mm)	151[f]
Average temperature (January and July, centigrade)	9.9/34.8[e]

a 1990. b 1988. c 1992. d 1991. e 1980. f 1987. g 1989.

Ireland

Region	Northern Europe
Location (longitude, latitude)	53°34'N 6°25'W
Currency	Irish pound
Population (1995 est., in thousands)	3469
Surface area (square kms)	70284
Population density (pop. per square km)	50
Sex ratio (females per 100 males)	100
Largest city (pop. in thousands)	Dublin (926)
UN membership date	14-Dec-1955
Major language(s)	Irish, English

Economic indicators	1985	1994
Exchange rate (US$)	0.80	0.64
Consumer price index (1980=100)	178	233[a]
Industrial production index (1980 = 100)	128	261[b]
Unemployment (%)	...	15.0[c]
Balance of payments, current account (million US$)	−690	2629[d]
Tourist arrivals (in thousands)	3007[e]	3666[d]
GDP (million US$)	18726	43593[f]
GDP (per capita US$)	5272	12480[f]
Long-term rate of change in GDP (% per annum)	3.1	2.5[f]
Gross fixed capital formation (% of GDP)	19.0	18.6[g]
Economically active female population (%)	29.7[i]	30.7
Economically active male population (%)	76.8[h]	75.7
Annual growth of econ. active female pop. (%)	2.0[g]	...
Annual growth of econ. active male pop. (%)	1.2[g]	...
Labour force in industry (%)	29.4[i]	28.6[f]
Labour force in agriculture (%)	16.9[i]	13.7[f]
Agricultural production index (1979-1981=100)	114[e]	132[d]
Food production index (1979-1981=100)	114[e]	132[d]
Commercial energy production (1,000 Mt coal equiv)	2723[h]	4733[f]
Motor vehicles (per 1,000 population)	229	272[g]
Telephone lines (per 100 inhabitants)	22.4[i]	30.0[f]

Largest export industries		Major trading partners			1992
	(% of exports)		(% of exports)	(% of imports)	
Metal manufacture	39	United Kingdom	32	United Kingdom	42
Chemicals	23	Germany	13	United States	14
Food, beverages, tobacco	23	France[k]	10	Germany	8

Social indicators	1990/95
Growth rate of population (% per annum)	−0.2
Age group 0-14 years (%)	24.7
Age group 60+ years (women and men, %)	8.6/6.9
Life expectancy at birth (women and men, years)	78/73
Infant mortality rate (per 1,000 births)	7
Total fertility rate (births per woman)	2
Urban population (%)	58
Urban population growth rate (% per annum)	0.3
Rural population growth rate (% per annum)	−0.8
Foreign-born (1985,%)	8.0
Refugees	500
Government education expenditure (% of GDP)	6
Primary-secondary gross enrolment ratio (f and m, per 100)	101/97[l]
Third-level students (per 100,000 population)	2308[l]
Newspaper circulation (per 1,000 population)	169
Television receivers (per 1,000 population)	76
Intentional homicides (1986, per 100,000 population)	1
Parliamentary seats (women and men, %)	12/88

Environment	1990/91
Threatened species	11
Forested area (%)	4.9
CO_2 emissions (10,000 Mt)	8798
Energy consumption (1,000 Mt coal equiv)	4143
Precipitation (mm)	758[j]
Average temperature (January and July, centigrade)	4.7/15.1[h]

a August 1994. b June 1994. c Unemployment insurance statistics. d 1992. e 1988. f 1991. g 1990. h 1980. i 1983. j 1987. k Includes Monaco. l 1989.

Israel

Region	Western Asia
Location (longitude, latitude)	32°07'N 34°79'E
Currency	new shekel
Population (1995 est., in thousands)	5884
Surface area (square kms)	21056
Population density (pop. per square km)	236
Sex ratio (females per 100 males)	102
Largest city (pop. in thousands)	Tel-Aviv (1902)
UN membership date	11-May-1949
Major language(s)	Hebrew, Arabic

Economic indicators	1985	1994
Exchange rate (US$)	1.50	3.01
Consumer price index (1980=100)	22500	111772
Industrial production index (1980 = 100)	120	184[a]
Unemployment (%)	...	7.4[b]
Balance of payments, current account (million US$)	1144	−1373[c]
Tourist arrivals (in thousands)	1170[d]	1502[e]
GDP (million US$)	25867	62684[f]
GDP (per capita US$)	6111	12869[f]
Long-term rate of change in GDP (% per annum)	3.6	7.1[f]
Gross fixed capital formation (% of GDP)	17.5	22.4[f]
Economically active female population (%)	36.7[g]	36.9
Economically active male population (%)	75.3[g]	75.1
Annual growth of econ. active female pop. (%)	3.2[h]	...
Annual growth of econ. active male pop. (%)	2.3[h]	...
Labour force in industry (%)	30.8[g]	28.5[e]
Labour force in agriculture (%)	6.3[g]	3.5[e]
Agricultural production index (1979-1981=100)	114[d]	110[e]
Food production index (1979-1981=100)	121[d]	126[e]
Commercial energy production (1,000 Mt coal equiv)	220[g]	50[f]
Motor vehicles (per 1,000 population)	172	209[h]
Telephone lines (per 100 inhabitants)	32.1[i]	34.9[f]

Largest export industries		Major trading partners			1992
	(% of exports)		(% of exports)		(% of imports)
Metal manufacture	33	United States	31	United States	17
Mining, quarrying	30	United Kingdom	8	Belgium[j]	13
Chemicals	16	Germany	6	Germany	12

Social indicators	1990/95
Growth rate of population (% per annum)	4.7
Age group 0-14 years (%)	28.8
Age group 60+ years (women and men, %)	7.4/5.7
Life expectancy at birth (women and men, years)	78/75
Infant mortality rate (per 1,000 births)	9
Total fertility rate (births per woman)	3
Urban population (%)	93
Urban population growth rate (% per annum)	4.9
Rural population growth rate (% per annum)	2.0
Foreign-born (1985,%)	33.9
Government education expenditure (% of GDP)	6[k]
Primary-secondary gross enrolment ratio (f and m, per 100)	92/88[k]
Third-level students (per 100,000 population)	2655[k]
Newspaper circulation (per 1,000 population)	258
Television receivers (per 1,000 population)	269
Intentional homicides (1986, per 100,000 population)	6
Parliamentary seats (women and men, %)	9/91

Environment	1990/91
Threatened species	30
Forested area (%)	5.4
CO_2 emissions (10,000 Mt)	9707
Energy consumption (1,000 Mt coal equiv)	2999
Precipitation (mm)	492[l]
Average temperature (January and July, centigrade)	8.6/23.3[g]

a June 1994. b Labour force sample surveys. c 1993. d 1988. e 1992.
f 1991. g 1980. h 1990. i 1987. j Includes Luxemburg. k 1989.
l Jerusalem.

Italy

Region	Southern Europe
Location (longitude, latitude)	45°45'N 9°18'E
Currency	Italian lire
Population (1995 est., in thousands)	57910
Surface area (square kms)	301268
Population density (pop. per square km)	189
Sex ratio (females per 100 males)	106
Largest city (pop. in thousands)	Milan (5279)
UN membership date	14-Dec-1955
Major language(s)	Italian

Economic indicators	1985	1994
Exchange rate (US$)	1678.50	1556.50
Consumer price index (1980=100)	190	305[a]
Industrial production index (1980 = 100)	97	123[a]
Unemployment (%)	...	11.6[b]
Balance of payments, current account (million US$)	−3865	−27994[c]
Tourist arrivals (in thousands)	26155[d]	26113[c]
GDP (million US$)	424512	1150526[e]
GDP (per capita US$)	7429	19930[e]
Long-term rate of change in GDP (% per annum)	2.6	1.4[e]
Gross fixed capital formation (% of GDP)	20.7	19.8[e]
Economically active female population (%)	29.9[f]	30.2
Economically active male population (%)	70.1[f]	68.6
Annual growth of econ. active female pop. (%)	1.1[g]	...
Annual growth of econ. active male pop. (%)	0.3[g]	...
Labour force in industry (%)	37.2[f]	32.0[e]
Labour force in agriculture (%)	14.0[f]	8.4[e]
Agricultural production index (1979-1981=100)	100[d]	105[c]
Food production index (1979-1981=100)	100[d]	105[c]
Commercial energy production (1,000 Mt coal equiv)	29246[f]	39774[e]
Motor vehicles (per 1,000 population)	434	538[e]
Telephone lines (per 100 inhabitants)	33.3[h]	40.0[e]

Largest export industries		Major trading partners			1992
	(% of exports)		(% of exports)		(% of imports)
Metal manufacture	47	Germany	20	Germany	22
Textiles	18	France[i]	15	France[i]	15
Chemicals	12	United States	7	Netherlands	6

Social indicators	1990/95
Growth rate of population (% per annum)	0.9
Age group 0-14 years (%)	15.4
Age group 60+ years (women and men, %)	12.3/9.0
Life expectancy at birth (women and men, years)	80/74
Infant mortality rate (per 1,000 births)	8
Total fertility rate (births per woman)	1
Urban population (%)	71
Urban population growth rate (% per annum)	0.6
Rural population growth rate (% per annum)	−1.0
Foreign-born (1985,%)	2.3
Refugees	12400
Government education expenditure (% of GDP)	3
Primary-secondary gross enrolment ratio (f and m, per 100)	84/83[i]
Third-level students (per 100,000 population)	2545
Newspaper circulation (per 1,000 population)	106[j]
Television receivers (per 1,000 population)	421
Intentional homicides (1986, per 100,000 population)	4
Parliamentary seats (women and men, %)	8/92

Environment	1990/91
Threatened species	76
Forested area (%)	22.4
CO_2 emissions (10,000 Mt)	109857[i]
Energy consumption (1,000 Mt coal equiv)	3998
Precipitation (mm)	903[h]
Average temperature (January and July, centigrade)	0.6/23.0[f]

a July 1994. b Labour force sample surveys. c 1992. d 1988. e 1991.
f 1980. g 1990. h 1987. i Includes Monaco. j 1989.

Jamaica

Region	Caribbean
Location (longitude, latitude)	18°00'N 76°77'W[a]
Currency	Jamaican dollar
Population (1995 est., in thousands)	2547
Surface area (square kms)	10990
Population density (pop. per square km)	215
Sex ratio (females per 100 males)	100
Largest city (pop. in thousands)	Kingston (638)[a]
UN membership date	18-Sep-1962
Major language(s)	English

Economic indicators	1985	1994
Exchange rate (US$)	5.48	33.28
Consumer price index (1980=100)	214	546[b]
Balance of payments, current account (million US$)	−304	117[c]
Tourist arrivals (in thousands)	649[d]	909[c]
GDP (million US$)	2015	3497[e]
GDP (per capita US$)	872	1431[e]
Long-term rate of change in GDP (% per annum)	−4.6	0.2[e]
Gross fixed capital formation (% of GDP)	23.0	29.1[f]
Economically active female population (%)	65.4[g]	68.3
Economically active male population (%)	83.0[g]	83.1
Annual growth of econ. active female pop. (%)	3.3[f]	...
Annual growth of econ. active male pop. (%)	2.6[f]	...
Labour force in industry (%)	15.2[g]	22.6[f]
Labour force in agriculture (%)	37.3[g]	26.1[f]
Agricultural production index (1979-1981=100)	106[d]	126[c]
Food production index (1979-1981=100)	106[d]	126[c]
Commercial energy production (1,000 Mt coal equiv)	15[g]	16[e]
Motor vehicles (per 1,000 population)	30	44[e]
Telephone lines (per 100 inhabitants)	3.5[h]	4.7[e]

Largest export industries	Major trading partners		1992	
(% of exports)	(% of exports)		(% of imports)	
...	United States	37	United States	53
...	United Kingdom	17	Mexico	7
...	Canada	11	Japan	5

Social indicators	1990/95
Growth rate of population (% per annum)	1.0
Age group 0-14 years (%)	30.9
Age group 60+ years (women and men, %)	4.8/4.0
Life expectancy at birth (women and men, years)	76/71
Infant mortality rate (per 1,000 births)	14
Total fertility rate (births per woman)	2
Contraceptive use (% of currently married women)	67
Urban population (%)	55
Urban population growth rate (% per annum)	2.2
Rural population growth rate (% per annum)	−0.3
Foreign-born (1985,%)	0.9
Government education expenditure (% of GDP)	4
Primary-secondary gross enrolment ratio (f and m, per 100)	82/78[i]
Third-level students (per 100,000 population)	515[i]
Newspaper circulation (per 1,000 population)	64
Television receivers (per 1,000 population)	131
Intentional homicides (1986, per 100,000 population)	19
Parliamentary seats (women and men, %)	12/88

Environment	1990/91
Threatened species	29
Forested area (%)	16.8
CO_2 emissions (10,000 Mt)	1275
Energy consumption (1,000 Mt coal equiv)	834
Precipitation (mm)	811[h]
Average temperature (January and July, centigrade)	25.4/28.3[g]

a Capital city. b December 1993. c 1992. d 1988. e 1991. f 1990. g 1980. h 1987. i 1989.

Japan

Region	Eastern Asia
Location (longitude, latitude)	35°68'N 139°77'E
Currency	yen
Population (1995 est., in thousands)	125879
Surface area (square kms)	377801
Population density (pop. per square km)	328
Sex ratio (females per 100 males)	103
Largest city (pop. in thousands)	Tokyo (25013)
UN membership date	18-Dec-1956
Major language(s)	Japanese

Economic indicators	1985	1994
Exchange rate (US$)	200.50	98.45
Consumer price index (1980=100)	114	131
Industrial production index (1980 = 100)	118	143[a]
Unemployment (%)	...	3.0[b]
Balance of payments, current account (million US$)	49	131[c]
Tourist arrivals (in thousands)	1116[d]	2103[e]
GDP (million US$)	1343251	3346411[f]
GDP (per capita US$)	11116	26983[f]
Long-term rate of change in GDP (% per annum)	5.0	4.1[f]
Gross fixed capital formation (% of GDP)	27.5	31.7[f]
Economically active female population (%)	47.3[g]	49.7
Economically active male population (%)	82.0[g]	78.3
Annual growth of econ. active female pop. (%)	0.6[h]	...
Annual growth of econ. active male pop. (%)	0.9[h]	...
Labour force in industry (%)	35.4[g]	34.6[e]
Labour force in agriculture (%)	10.4[g]	6.4[e]
Agricultural production index (1979-1981=100)	97[d]	96[e]
Food production index (1979-1981=100)	100[d]	99[e]
Commercial energy production (1,000 Mt coal equiv)	62939[g]	104816[f]
Motor vehicles (per 1,000 population)	374	475[f]
Telephone lines (per 100 inhabitants)	39.7[i]	45.4[e]

Largest export industries		Major trading partners			1992
	(% of exports)		(% of exports)		(% of imports)
Metal manufacture	78	United States	28	United States	23
Chemicals	9	Hong Kong	6	China	7
Basic metal industry	5	Germany	6	Indonesia	5

Social indicators	1990/95
Growth rate of population (% per annum)	0.4
Age group 0-14 years (%)	16.8
Age group 60+ years (women and men, %)	11.1/8.7
Life expectancy at birth (women and men, years)	82/76
Infant mortality rate (per 1,000 births)	5
Total fertility rate (births per woman)	2
Contraceptive use (% of currently married women)	64
Urban population (%)	78
Urban population growth rate (% per annum)	0.6
Rural population growth rate (% per annum)	−0.3
Foreign-born (1985,%)	0.6
Refugees	8200
Government education expenditure (% of GDP)	5[d]
Primary-secondary gross enrolment ratio (f and m, per 100)	99/98[j]
Third-level students (per 100,000 population)	2184[j]
Newspaper circulation (per 1,000 population)	587
Television receivers (per 1,000 population)	613
Intentional homicides (1986, per 100,000 population)	1
Parliamentary seats (women and men, %)	2/98

Environment	1990/91
Threatened species	151
Forested area (%)	66.5
CO_2 emissions (10,000 Mt)	297802
Energy consumption (1,000 Mt coal equiv)	4754
Precipitation (mm)	1563[i]
Average temperature (January and July, centigrade)	3.7/25.1[g]

a August 1994. b Labour force sample surveys. c 1993. d 1988. e 1992. f 1991. g 1980. h 1990. i 1987. j 1989.

Jordan

Region	Western Asia
Location (longitude, latitude)	31°95'N 35°94'E
Currency	Jordanian dinar
Population (1995 est., in thousands)	4755
Surface area (square kms)	97740
Population density (pop. per square km)	42
Sex ratio (females per 100 males)	95
Largest city (pop. in thousands)	Amman (955)
UN membership date	14-Dec-1955
Major language(s)	Arabic

Economic indicators	1985	1994
Exchange rate (US$)	0.37	0.70
Consumer price index (1980=100)	130	241[a]
Industrial production index (1980 = 100)	155	178[b]
Balance of payments, current account (million US$)	−261	−765[b]
Tourist arrivals (in thousands)	608[c]	661[b]
GDP (million US$)	4818	4041[d]
GDP (per capita US$)	1414	974[d]
Long-term rate of change in GDP (% per annum)	3.5	0.5[d]
Gross fixed capital formation (% of GDP)	24.0	20.0[e]
Economically active female population (%)	6.8[f]	10.3
Economically active male population (%)	76.1[f]	77.8
Annual growth of econ. active female pop. (%)	5.2[e]	...
Annual growth of econ. active male pop. (%)	2.6[e]	...
Labour force in industry (%)	21.0[f]	22.8[d]
Agricultural production index (1979-1981=100)	167[c]	214[b]
Food production index (1979-1981=100)	169[c]	217[b]
Commercial energy production (1,000 Mt coal equiv)	25[e]	11[d]
Motor vehicles (per 1,000 population)	58	57[d]
Telephone lines (per 100 inhabitants)	5.6[g]	6.4[d]

Largest export industries		Major trading partners			1992
	(% of exports)		(% of exports)		(% of imports)
Chemicals	30	India	12	Iraq	13
Mining, quarrying	26	Saudi Arabia	9	United States	11
Metal manufacture	15	Iraq	9	Germany	9

Social indicators	1990/95
Growth rate of population (% per annum)	3.4
Age group 0-14 years (%)	43.6
Age group 60+ years (women and men, %)	2.3/2.1
Life expectancy at birth (women and men, years)	70/66
Infant mortality rate (per 1,000 births)	36
Total fertility rate (births per woman)	6
Contraceptive use (% of currently married women)	35
Urban population (%)	71
Urban population growth rate (% per annum)	4.4
Rural population growth rate (% per annum)	1.1
Foreign-born (1985,%)	26.5
Refugees	300
Government education expenditure (% of GDP)	4
Primary-secondary gross enrolment ratio (f and m, per 100)	93/91[h]
Third-level students (per 100,000 population)	2006
Newspaper circulation (per 1,000 population)	56
Television receivers (per 1,000 population)	80
Intentional homicides (1986, per 100,000 population)	2
Parliamentary seats (women and men, %)	0/100

Environment	1990/91
Threatened species	16
Forested area (%)	0.8
CO_2 emissions (10,000 Mt)	2732
Energy consumption (1,000 Mt coal equiv)	1030
Precipitation (mm)	273[g]
Average temperature (January and July, centigrade)	8.2/25.2[f]

a August 1994. b 1992. c 1988. d 1991. e 1990. f 1980. g 1987. h 1989.

Kazakhstan

Region	Central Asia
Location (longitude, latitude)	43°14'N 76°56'E
Currency	rouble
Surface area (square kms)	2717300
Largest city (pop. in thousands)	Alma-Ata (1158)
UN membership date	02-Mar-1992
Major language(s)	Kazakh, Russian

Economic indicators	1985	1994
Exchange rate (US$)	...	12.00
GDP (million US$)	37103	46445[a]
GDP (per capita US$)	2344	2749[a]
Long-term rate of change in GDP (% per annum)	6.2	−8.6[a]
Economically active female population (%)	54.2[b]	54.1
Economically active male population (%)	79.2[b]	80.2
Telephone lines (per 100 inhabitants)	8.6[c]	11.1[a]

Social indicators	1990/95
Contraceptive use (% of currently married women)	30

Environment	1990/91
Threatened species	49

a 1991. b 1980. c 1987.

Kenya

Region	Eastern Africa
Location (longitude, latitude)	1°28'S 36°82'E
Currency	Kenyan shilling
Population (1995 est., in thousands)	27885
Surface area (square kms)	580367
Population density (pop. per square km)	45
Sex ratio (females per 100 males)	100
Largest city (pop. in thousands)	Nairobi (1518)
UN membership date	16-Dec-1963
Major language(s)	English, Swahili, Kuyu, Luo, Kamba, Luhya, Nandi-kipsigis, Kisii

Economic indicators	1985	1994
Exchange rate (US$)	16.28	48.01
Consumer price index (1980=100)	187	305[a]
Balance of payments, current account (million US$)	–113	–98[b]
Tourist arrivals (in thousands)	695[c]	699[b]
GDP (million US$)	6138	8261[b]
GDP (per capita US$)	309	339[d]
Long-term rate of change in GDP (% per annum)	4.3	1.7[d]
Gross fixed capital formation (% of GDP)	17.4	18.8[d]
Economically active female population (%)	62.6[e]	55.4
Economically active male population (%)	90.7[e]	89.1
Annual growth of econ. active female pop. (%)	3.3[f]	...
Annual growth of econ. active male pop. (%)	3.9[f]	...
Labour force in industry (%)	21.6[e]	20.0[d]
Labour force in agriculture (%)	23.0[e]	18.9[d]
Agricultural production index (1979-1981=100)	146[c]	135[b]
Food production index (1979-1981=100)	146[c]	138[b]
Commercial energy production (1,000 Mt coal equiv)	130[e]	706[d]
Motor vehicles (per 1,000 population)	12	13[d]
Telephone lines (per 100 inhabitants)	0.7[g]	0.8[d]

Social indicators	1990/95
Growth rate of population (% per annum)	3.3
Age group 0-14 years (%)	47.4
Age group 60+ years (women and men, %)	2.4/2.0
Life expectancy at birth (women and men, years)	61/57
Infant mortality rate (per 1,000 births)	66
Total fertility rate (births per woman)	6
Contraceptive use (% of currently married women)	33
Urban population (%)	28
Urban population growth rate (% per annum)	6.6
Rural population growth rate (% per annum)	2.2
Foreign-born (1985,%)	0.8
Refugees	401900
Government education expenditure (% of GDP)	6
Primary-secondary gross enrolment ratio (f and m, per 100)	74/79
Third-level students (per 100,000 population)	135[h]
Newspaper circulation (per 1,000 population)	15
Television receivers (per 1,000 population)	10
Parliamentary seats (women and men, %)	3/97

Environment	1990/91
Threatened species	47
Forested area (%)	4.0
CO_2 emissions (10,000 Mt)	1323
Energy consumption (1,000 Mt coal equiv)	109
Precipitation (mm)	926[g]
Average temperature (January and July, centigrade)	17.8/14.9[e]

a April 1994. b 1992. c 1988. d 1991. e 1980. f 1990. g 1987. h 1989.

Kiribati

Region Oceania-Micronesia
Location (longitude, latitude) 1°25'N 173°20'E[a]
Currency Australian dollar
Population (1995 est., in thousands) 56
Surface area (square kms) 726
Population density (pop. per square km) 91
Sex ratio (females per 100 males) 103
Largest city (pop. in thousands) Tarawa (25)[a]
Major language(s) English, Kiribati

Economic indicators	1985	1994
Exchange rate (US$)	1.47	1.35
Consumer price index (1980=100)	131	181
Balance of payments, current account (million US$)	6	12[b]
Tourist arrivals (in thousands)	3[c]	4[b]
GDP (million US$)	20	39[d]
GDP (per capita US$)	306	544[d]
Long-term rate of change in GDP (% per annum)	−9.3	3.9[d]
Telephone lines (per 100 inhabitants)	1.4[f]	1.8[d]

Largest export industries	Major trading partners			1992
(% of exports)	(% of exports)		(% of imports)	
Agriculture 83	Bangladesh	67	Australia	38
Other manufacturing 16	United States	12	Japan	23
Basic metal industry 2	Australia	5	Fiji	11

Social indicators	1990/95
Age group 0-14 years (%)	41.1[e]
Age group 60+ years (women and men, %)	3.3/2.5[g]
Contraceptive use (% of currently married women)	37[ch]
Urban population (%)	39
Urban population growth rate (% per annum)	3.8
Rural population growth rate (% per annum)	1.2
Foreign-born (1985,%)	4.0
Intentional homicides (1986, per 100,000 population)	32
Parliamentary seats (women and men, %)	0/100

Environment	1990/91
Threatened species	7
Forested area (%)	2.8
CO_2 emissions (10,000 Mt)	6
Energy consumption (1,000 Mt coal equiv)	139
Average temperature (January and July, centigrade)	27.7/27.8[e]

a Capital city. b 1992. c 1988. d 1991. e 1980. f 1987. g 1978. h Source: NatSTB.

Korea, Democratic People's Republic of

Region Eastern Asia
Location (longitude, latitude) 39°01'N 125°74'E
Currency won
Population (1995 est., in thousands) 23922
Sex ratio (females per 100 males) 103
Largest city (pop. in thousands) Pyongyang (2230)
UN membership date 17-Sep-1991
Major language(s) Korean

Economic indicators	1985	1994
GDP (million US$)	16700	21310[a]
GDP (per capita US$)	840	960[a]
Long-term rate of change in GDP (% per annum)	9.6	0.0[a]
Economically active female population (%)	63.2[b]	66.2
Economically active male population (%)	81.2[b]	83.3
Annual growth of econ. active female pop. (%)	3.1[c]	...
Annual growth of econ. active male pop. (%)	2.9[c]	...
Agricultural production index (1979-1981=100)	126[d]	115[e]
Food production index (1979-1981=100)	125[d]	114[e]
Commercial energy production (1,000 Mt coal equiv)	44764[b]	84100[a]
Telephone lines (per 100 inhabitants)	3.3[f]	3.6[a]

Social indicators	1990/95
Growth rate of population (% per annum)	1.9
Age group 0-14 years (%)	29.1
Age group 60+ years (women and men, %)	4.5/2.5
Life expectancy at birth (women and men, years)	74/68
Infant mortality rate (per 1,000 births)	24
Total fertility rate (births per woman)	2
Urban population (%)	61
Urban population growth rate (% per annum)	2.4
Rural population growth rate (% per annum)	1.2
Foreign-born (1985, %)	0.2
Newspaper circulation (per 1,000 population)	230
Television receivers (per 1,000 population)	15
Parliamentary seats (women and men, %)	20/80

Environment	1990/91
Threatened species	30
Forested area (%)	74.4
CO_2 emissions (10,000 Mt)	66385
Energy consumption (1,000 Mt coal equiv)	4196

a 1991. b 1980. c 1990. d 1988. e 1992. f 1987.

Korea, Republic of

Region	Eastern Asia
Location (longitude, latitude)	37°58'N 126°98'E
Currency	won
Population (1995 est., in thousands)	45182
Surface area (square kms)	99016
Population density (pop. per square km)	437
Sex ratio (females per 100 males)	98
Largest city (pop. in thousands)	Seoul (10979)
UN membership date	17-Sep-1991
Major language(s)	Korean

Economic indicators	1985	1994
Exchange rate (US$)	890.20	798.90
Consumer price index (1980=100)	141	240
Industrial production index (1980 = 100)	165	392[a]
Unemployment (%)	...	2.2[b]
Balance of payments, current account (million US$)	−887	−4529[c]
Tourist arrivals (in thousands)	2340[d]	3231[c]
GDP (million US$)	92925	282971[e]
GDP (per capita US$)	2277	6462[e]
Long-term rate of change in GDP (% per annum)	6.9	8.4[e]
Gross fixed capital formation (% of GDP)	28.2	38.0[e]
Economically active female population (%)	39.5[f]	40.9
Economically active male population (%)	76.0[f]	78.8
Annual growth of econ. active female pop. (%)	2.9[g]	...
Annual growth of econ. active male pop. (%)	2.6[g]	...
Labour force in industry (%)	29.0[f]	34.6[c]
Labour force in agriculture (%)	34.0[f]	16.0[c]
Agricultural production index (1979-1981=100)	114[f]	113[c]
Food production index (1979-1981=100)	116[d]	114[c]
Commercial energy production (1,000 Mt coal equiv)	13774[f]	31218[e]
Motor vehicles (per 1,000 population)	27	97[e]
Telephone lines (per 100 inhabitants)	20.6[h]	33.3[e]

Largest export industries		Major trading partners			1992
	(% of exports)		(% of exports)		(% of imports)
Metal manufacture	48	United States	24	Japan	24
Textiles	26	Japan	15	United States	23
Chemicals	12	Hong Kong	8	Saudi Arabia	5

Social indicators	1990/95
Growth rate of population (% per annum)	0.8
Age group 0-14 years (%)	23.3
Age group 60+ years (women and men, %)	5.3/3.5
Life expectancy at birth (women and men, years)	74/68
Infant mortality rate (per 1,000 births)	21
Total fertility rate (births per woman)	2
Contraceptive use (% of currently married women)	79
Urban population (%)	78
Urban population growth rate (% per annum)	2.3
Rural population growth rate (% per annum)	−3.6
Foreign-born (1985,%)	2.1
Refugees	100
Government education expenditure (% of GDP)	3
Primary-secondary gross enrolment ratio (f and m, per 100)	97/97
Third-level students (per 100,000 population)	3953
Newspaper circulation (per 1,000 population)	277
Television receivers (per 1,000 population)	208
Intentional homicides (1986, per 100,000 population)	1
Parliamentary seats (women and men, %)	1/99

Environment	1990/91
Threatened species	29
Forested area (%)	65.4
CO_2 emissions (10,000 Mt)	72229
Energy consumption (1,000 Mt coal equiv)	2977
Precipitation (mm)	1093[i]
Average temperature (January and July, centigrade)	−4.0/23.9[i]

a July 1994. b Labour force sample surveys. c 1992. d 1988. e 1991.
f 1980. g 1990. h 1987. i Inchon.

Kuwait

Region	Western Asia
Location (longitude, latitude)	29°30'N 47°45'E
Currency	Kuwaiti dinar
Population (1995 est., in thousands)	1604
Surface area (square kms)	17818
Population density (pop. per square km)	118
Sex ratio (females per 100 males)	97
Largest city (pop. in thousands)	Kuwait City (1090)
UN membership date	14-May-1963
Major language(s)	Arabic

Economic indicators	1985	1994
Exchange rate (US$)	0.29	0.30
Consumer price index (1980=100)	124	135[a]
Balance of payments, current account (million US$)	5150	−873[b]
Tourist arrivals (in thousands)	80[c]	65[b]
GDP (million US$)	21429	24235[d]
GDP (per capita US$)	12458	11618[d]
Long-term rate of change in GDP (% per annum)	−4.3	−44.0[d]
Gross fixed capital formation (% of GDP)	19.8	...
Economically active female population (%)	20.3[e]	27.2
Economically active male population (%)	85.8[e]	82.9
Annual growth of econ. active female pop. (%)	10.0[a]	...
Annual growth of econ. active male pop. (%)	5.9[a]	...
Commercial energy production (1,000 Mt coal equiv)[f]	134712[e]	14858[d]
Motor vehicles (per 1,000 population)	324	...
Telephone lines (per 100 inhabitants)	13.1[g]	16.1[d]

Social indicators	1990/95
Growth rate of population (% per annum)	−5.8
Age group 0-14 years (%)	41.1
Age group 60+ years (women and men, %)	1.2/1.8
Life expectancy at birth (women and men, years)	78/73
Infant mortality rate (per 1,000 births)	14
Total fertility rate (births per woman)	4
Contraceptive use (% of currently married women)	35[g]
Urban population (%)	97
Urban population growth rate (% per annum)	−5.6
Rural population growth rate (% per annum)	−12.3
Foreign-born (1985, %)	59.4
Refugees	124900
Government education expenditure (% of GDP)	5[h]
Primary-secondary gross enrolment ratio (f and m, per 100)	86/89[i]
Third-level students (per 100,000 population)	1384[c]
Newspaper circulation (per 1,000 population)	210
Television receivers (per 1,000 population)	283
Intentional homicides (1986, per 100,000 population)	1
Parliamentary seats (women and men, %)	0/100

Environment	1990/91
Threatened species	10
Forested area (%)	0.1
CO_2 emissions (10,000 Mt)	3232[f]
Energy consumption (1,000 Mt coal equiv)[f]	1925
Precipitation (mm)	111[j]
Average temperature (January and July, centigrade)	13.9/36.7[e]

a 1990. b 1992. c 1988. d 1991. e 1980. f Includes part of the Neutral Zone. g 1987. h 1986. i 1989. j Shuwaikh.

Kyrgyzstan

Region	Central Asia
Location (longitude, latitude)	42°53'N 76°46'E[a]
Currency	rouble
Surface area (square kms)	198500
Largest city (pop. in thousands)	Bishkek (629)[a]
UN membership date	02-Mar-1992
Major language(s)	Kirghiz, Russian, Uzbec

Economic indicators	1985	1994
Exchange rate (US$)	...	8.80
GDP (million US$)	7301	8945[b]
GDP (per capita US$)	1819	2007[b]
Long-term rate of change in GDP (% per annum)	−5.3	−10.4[b]
Economically active female population (%)	55.6[c]	57.7
Economically active male population (%)	76.1[c]	77.9
Telephone lines (per 100 inhabitants)	5.8[d]	7.3[b]

Social indicators	1990/95
Contraceptive use (% of currently married women)	31

Environment	1990/91
Threatened species	17

a Capital city. b 1991. c 1980. d 1987.

Lao People's Democratic Republic

Region	South-eastern Asia
Location (longitude, latitude)	17°57'N 102°34'E[a]
Currency	new kip
Population (1995 est., in thousands)	4882
Surface area (square kms)	236800
Population density (pop. per square km)	18
Sex ratio (females per 100 males)	103
Largest city (pop. in thousands)	Vientiane (415)[a]
UN membership date	14-Dec-1955
Major language(s)	Lao, Thai, Kmhu

Economic indicators	1985	1994
Exchange rate (US$)	95.00	718.00
Balance of payments, current account (million US$)	−94	−41[b]
Tourist arrivals (in thousands)	25[c]	25[b]
GDP (million US$)	2496	1042[d]
GDP (per capita US$)	695	240[d]
Long-term rate of change in GDP (% per annum)	5.1	5.0[d]
Economically active female population (%)	78.0[e]	68.2
Economically active male population (%)	89.9[e]	88.4
Annual growth of econ. active female pop. (%)	1.6[f]	...
Annual growth of econ. active male pop. (%)	2.0[f]	...
Agricultural production index (1979-1981=100)	158[c]	160[b]
Food production index (1979-1981=100)	158[c]	160[b]
Commercial energy production (1,000 Mt coal equiv)	114[e]	115[d]
Telephone lines (per 100 inhabitants)	0.2[g]	0.1[d]

Social indicators	1990/95
Growth rate of population (% per annum)	3.0
Age group 0-14 years (%)	44.8
Age group 60+ years (women and men, %)	2.6/2.2
Life expectancy at birth (women and men, years)	53/50
Infant mortality rate (per 1,000 births)	97
Total fertility rate (births per woman)	7
Urban population (%)	22
Urban population growth rate (% per annum)	6.1
Rural population growth rate (% per annum)	2.2
Foreign-born (1985,%)	0.4
Government education expenditure (% of GDP)	1[c]
Primary-secondary gross enrolment ratio (f and m, per 100)	51/70
Third-level students (per 100,000 population)	118[h]
Newspaper circulation (per 1,000 population)	3
Television receivers (per 1,000 population)	6
Parliamentary seats (women and men, %)	9/91

Environment	1990/91
Threatened species	65
Forested area (%)	53.6
CO_2 emissions (10,000 Mt)	69
Energy consumption (1,000 Mt coal equiv)	38
Precipitation (mm)	1683[g]
Average temperature (January and July, centigrade)	21.5/27.7[e]

a Capital city. b 1992. c 1988. d 1991. e 1980. f 1990. g 1987. h 1989.

Latvia

Region Northern Europe
Location (longitude, latitude) 56°93'N 24°10'E
Currency rouble
Surface area (square kms) 64500
Largest city (pop. in thousands) Riga (921)
UN membership date 17-Sep-1991
Major language(s) Latvian, Russian

Economic indicators	1985	1994
Exchange rate (US$)	...	0.60
Consumer price index (1980=100)	...	5556[a]
GDP (million US$)	9934	16436[b]
GDP (per capita US$)	3802	6119[b]
Long-term rate of change in GDP (% per annum)	−0.2	−8.3[b]
Gross fixed capital formation (% of GDP)	6.2[b]	...
Economically active female population (%)	58.9[c]	57.7
Economically active male population (%)	78.5[c]	79.5
Motor vehicles (per 1,000 population)	110	153[b]
Telephone lines (per 100 inhabitants)	21.5[d]	23.9

Social indicators	1990/95
Urban population (%)	73
Urban population growth rate (% per annum)	0.2
Rural population growth rate (% per annum)	−1.4
Television receivers (per 1,000 population)	372
Intentional homicides (1986, per 100,000 population)	9[e]

Environment	1990/91
Threatened species	33

a February 1994. b 1991. c 1980. d 1987. e Source: DYB92.

Lebanon

Region	Western Asia
Location (longitude, latitude)	33°90'N 35°50'E
Currency	Lebanese pound
Population (1995 est., in thousands)	3028
Surface area (square kms)	10400
Population density (pop. per square km)	264
Sex ratio (females per 100 males)	105
Largest city (pop. in thousands)	Beirut (1563)
UN membership date	24-Oct-1945
Major language(s)	Arabic

Economic indicators	1985	1994
Exchange rate (US$)	18.10	1666.00
GDP (million US$)	2071	1706[a]
GDP (per capita US$)	776	613[a]
Long-term rate of change in GDP (% per annum)	−16.8	3.0[a]
Economically active female population (%)	19.7[b]	25.0
Economically active male population (%)	73.0[b]	73.9
Annual growth of econ. active female pop. (%)	3.8[c]	...
Annual growth of econ. active male pop. (%)	1.2[c]	...
Agricultural production index (1979-1981=100)	138[d]	196[e]
Food production index (1979-1981=100)	141[d]	202[e]
Commercial energy production (1,000 Mt coal equiv)	104[b]	69[a]
Telephone lines (per 100 inhabitants)	13.7[c]	11.1[a]

Social indicators	1990/95
Growth rate of population (% per annum)	2.0
Age group 0-14 years (%)	34.2
Age group 60+ years (women and men, %)	4.5/3.8
Life expectancy at birth (women and men, years)	71/67
Infant mortality rate (per 1,000 births)	34
Total fertility rate (births per woman)	3
Urban population (%)	87
Urban population growth rate (% per annum)	2.8
Rural population growth rate (% per annum)	−2.7
Foreign-born (1985, %)	10.4
Refugees	6000
Primary-secondary gross enrolment ratio (f and m, per 100)	83/84[d]
Third-level students (per 100,000 population)	3071
Newspaper circulation (per 1,000 population)	117
Television receivers (per 1,000 population)	325
Parliamentary seats (women and men, %)	2/98

Environment	1990/91
Threatened species	13
Forested area (%)	7.7
CO_2 emissions (10,000 Mt)	2282
Energy consumption (1,000 Mt coal equiv)	1370
Precipitation (mm)	517[f]
Average temperature (January and July, centigrade)	13.9/26.2[b]

a 1991. b 1980. c 1990. d 1988. e 1992. f 1987.

Lesotho

Region Western Africa
Location (longitude, latitude) 29°28'S 27°30'E[a]
Currency maloti
Population (1995 est., in thousands) 1977
Surface area (square kms) 30355
Population density (pop. per square km) 60
Sex ratio (females per 100 males) 105
Largest city (pop. in thousands) Maseru (170)[a]
UN membership date 17-Oct-1966
Major language(s) English, Sotho, Zulu

Economic indicators	1985	1994
Exchange rate (US$)	2.56	3.56
Consumer price index (1980=100)	186	484
Balance of payments, current account (million US$)	–12	38[b]
Tourist arrivals (in thousands)	110[c]	155[b]
GDP (million US$)	252	596[d]
GDP (per capita US$)	163	333[d]
Long-term rate of change in GDP (% per annum)	3.5	0.0[d]
Gross fixed capital formation (% of GDP)	49.6	77.2[e]
Economically active female population (%)	70.9[f]	62.5
Economically active male population (%)	91.8[f]	90.4
Annual growth of econ. active female pop. (%)	1.5[e]	...
Annual growth of econ. active male pop. (%)	2.5[e]	...
Agricultural production index (1979-1981=100)	114[c]	81[b]
Food production index (1979-1981=100)	116[c]	79[b]
Motor vehicles (per 1,000 population)	13	...
Telephone lines (per 100 inhabitants)	0.7[g]	0.6[d]

Social indicators	1990/95
Growth rate of population (% per annum)	2.5
Age group 0-14 years (%)	40.7
Age group 60+ years (women and men, %)	3.4/2.6
Life expectancy at birth (women and men, years)	63/58
Infant mortality rate (per 1,000 births)	79
Total fertility rate (births per woman)	5
Contraceptive use (% of currently married women)	23
Urban population (%)	23
Urban population growth rate (% per annum)	6.0
Rural population growth rate (% per annum)	1.5
Foreign-born (1985,%)	1.3
Refugees	100
Government education expenditure (% of GDP)	5
Primary-secondary gross enrolment ratio (f and m, per 100)	84/70
Third-level students (per 100,000 population)	333[c]
Newspaper circulation (per 1,000 population)	11
Television receivers (per 1,000 population)	6
Parliamentary seats (women and men, %)	2/98

Environment	1990/91
Threatened species	12

a Capital city. b 1992. c 1988. d 1991. e 1990. f 1980. g 1987.

Liberia

Region Western Africa
Location (longitude, latitude) 6°20'N 10°46'W[a]
Currency Liberian dollar
Population (1995 est., in thousands) 3039
Surface area (square kms) 111369
Population density (pop. per square km) 24
Sex ratio (females per 100 males) 98
Largest city (pop. in thousands) Monrovia (668)[a]
UN membership date 02-Nov-1945
Major language(s) English, Kpelle, Bassa, Grebo, Dan, Mano, Loma

Economic indicators	1985	1994
Exchange rate (US$)	1.00[b]	1.00[b]
Consumer price index (1980=100)	119	162[c]
Balance of payments, current account (million US$)	56	...
GDP (million US$)	1055	1037[d]
GDP (per capita US$)	480	390[d]
Long-term rate of change in GDP (% per annum)	−2.0	−10.0[d]
Gross fixed capital formation (% of GDP)	12.0	...
Economically active female population (%)	40.6[e]	35.5
Economically active male population (%)	89.4[e]	87.2
Annual growth of econ. active female pop. (%)	2.2[c]	...
Annual growth of econ. active male pop. (%)	2.6[c]	...
Agricultural production index (1979-1981=100)	119[f]	76[g]
Food production index (1979-1981=100)	122[f]	91[g]
Commercial energy production (1,000 Mt coal equiv)	41[e]	20[d]
Motor vehicles (per 1,000 population)	14	...
Telephone lines (per 100 inhabitants)	0.3[h]	0.1[d]

Social indicators	1990/95
Growth rate of population (% per annum)	3.3
Age group 0-14 years (%)	46.0
Age group 60+ years (women and men, %)	2.8/2.6
Life expectancy at birth (women and men, years)	57/54
Infant mortality rate (per 1,000 births)	126
Total fertility rate (births per woman)	7
Contraceptive use (% of currently married women)	6[i]
Urban population (%)	51
Urban population growth rate (% per annum)	5.5
Rural population growth rate (% per annum)	1.3
Foreign-born (1985,%)	4.6
Refugees	100000
Government education expenditure (% of GDP)	5[j]
Primary-secondary gross enrolment ratio (f and m, per 100)	25/48[e]
Third-level students (per 100,000 population)	220
Newspaper circulation (per 1,000 population)	14
Television receivers (per 1,000 population)	18
Intentional homicides (1986, per 100,000 population)	2
Parliamentary seats (women and men, %)	6/94

Environment	1990/91
Threatened species	34
Forested area (%)	17.8
CO_2 emissions (10,000 Mt)	75
Energy consumption (1,000 Mt coal equiv)	57

a Capital city. b Fixed rate. c 1990. d 1991. e 1980. f 1988. g 1992. h 1987. i 1986. j 1985.

Libyan Arab Jamahiriya

Region	Northern Africa
Location (longitude, latitude)	32°88'N 13°19'E
Currency	Libyan dinar
Population (1995 est., in thousands)	5407
Surface area (square kms)	1759540
Population density (pop. per square km)	3
Sex ratio (females per 100 males)	92
Largest city (pop. in thousands)	Tripoli (2595)
UN membership date	14-Dec-1955
Major language(s)	Arabic

Economic indicators	1985	1994
Exchange rate (US$)	0.30	0.30
Balance of payments, current account (million US$)	1906	2201[a]
Tourist arrivals (in thousands)	98[b]	89[c]
GDP (million US$)	27963	44967[d]
GDP (per capita US$)	7386	9551[d]
Long-term rate of change in GDP (% per annum)	8.3	5.0[d]
Economically active female population (%)	7.2[e]	9.4
Economically active male population (%)	80.2[e]	76.8
Annual growth of econ. active female pop. (%)	5.9[a]	...
Annual growth of econ. active male pop. (%)	3.6[a]	...
Agricultural production index (1979-1981=100)	106[b]	186[c]
Food production index (1979-1981=100)	106[b]	187[c]
Commercial energy production (1,000 Mt coal equiv)	132809[e]	116564[d]
Motor vehicles (per 1,000 population)	166	169[a]
Telephone lines (per 100 inhabitants)	4.2[f]	5.7[d]

Social indicators	1990/95
Growth rate of population (% per annum)	3.5
Age group 0-14 years (%)	45.4
Age group 60+ years (women and men, %)	2.0/2.4
Life expectancy at birth (women and men, years)	65/62
Infant mortality rate (per 1,000 births)	68
Total fertility rate (births per woman)	6
Urban population (%)	86
Urban population growth rate (% per annum)	4.3
Rural population growth rate (% per annum)	-1.1
Foreign-born (1985,%)	14.3
Government education expenditure (% of GDP)	8[g]
Third-level students (per 100,000 population)	1548
Newspaper circulation (per 1,000 population)	15
Television receivers (per 1,000 population)	99

Environment	1990/91
Threatened species	17
Forested area (%)	0.4
CO_2 emissions (10,000 Mt)	11738
Energy consumption (1,000 Mt coal equiv)	4328
Precipitation (mm)	253[h]
Average temperature (January and July, centigrade)	10.9/27.1[e]

a 1990. b 1988. c 1992. d 1991. e 1980. f 1987. g 1986. h Idris.

Liechtenstein

Region Western Europe
Location (longitude, latitude) 47°14'N 9°52'E[a]
Currency Swiss franc
Surface area (square kms) 160
Population density (pop. per square km) 175
Largest city (pop. in thousands) Vaduz (6)[a]
UN membership date 18-Sep-1990
Major language(s) German

Economic indicators	1985	1994
Tourist arrivals (in thousands)	72[b]	72[c]
GDP (million US$)	529	1504[d]
GDP (per capita US$)	19596	53724[d]
Long-term rate of change in GDP (% per annum)	4.2	0.0[d]
Telephone lines (per 100 inhabitants)	51.2[e]	62.5[d]

Social indicators	1990/95
Urban population (%)	21
Urban population growth rate (% per annum)	1.3
Rural population growth rate (% per annum)	−0.1
Foreign-born (1985,%)	35.1
Newspaper circulation (per 1,000 population)	307
Television receivers (per 1,000 population)	345

Environment	1990/91
Threatened species	10
Forested area (%)	18.8

a Capital city. b 1988. c 1992. d 1991. e 1987.

Lithuania

Region Northern Europe
Location (longitude, latitude) 54°41'N 25°19'E[a]
Currency rouble
Surface area (square kms) 65200
Population density (pop. per square km) 57
Largest city (pop. in thousands) Vilnius (598)[a]
UN membership date 17-Sep-1991
Major language(s) Lithuanian, Russian, Polish

Economic indicators	1985	1994
Exchange rate (US$)	...	4.00
Consumer price index (1980=100)	106	3546
GDP (million US$)	13569	21802[b]
GDP (per capita US$)	3783	5820[b]
Long-term rate of change in GDP (% per annum)	9.6	−13.5[b]
Economically active female population (%)	57.6[c]	56.3
Economically active male population (%)	77.9[c]	79.6
Motor vehicles (per 1,000 population)	123	172[b]
Telephone lines (per 100 inhabitants)	17.4[d]	21.6[b]

Largest export industries	Major trading partners			1992
(% of exports)	(% of exports)		(% of imports)	
...	Russian Fed.	28	Russian Fed.	58
...	Ukraine	16	Ukraine	8
...	Belarus	12	Belarus	6

Social indicators	1990/95
Urban population (%)	72
Urban population growth rate (% per annum)	1.2
Rural population growth rate (% per annum)	−2.0
Television receivers (per 1,000 population)	374
Intentional homicides (1986, per 100,000 population)	8[e]

Environment	1990/91
Threatened species	33

a Capital city. b 1991. c 1980. d 1987. e Source: DYB92.

Luxembourg

Region Western Europe
Location (longitude, latitude) 49°61'N 6°12'E[a]
Currency Luxembourg franc
Population (1995 est., in thousands) 386
Surface area (square kms) 2586
Population density (pop. per square km) 144
Sex ratio (females per 100 males) 103
Largest city (pop. in thousands) Luxembourg-ville (83)[a]
UN membership date 24-Oct-1945
Major language(s) French, Luxemburgish

Economic indicators	1985	1994
Exchange rate (US$)	50.36	31.83
Consumer price index (1980=100)	140	113
Industrial production index (1980 = 100)	121	140[b]
Tourist arrivals (in thousands)	760[c]	796[d]
GDP (million US$)	3457	9336[e]
GDP (per capita US$)	9419	24896[e]
Long-term rate of change in GDP (% per annum)	2.9	3.1[e]
Gross fixed capital formation (% of GDP)	17.7	29.0[e]
Economically active female population (%)	32.5[f]	32.2
Economically active male population (%)	72.4[f]	72.8
Annual growth of econ. active female pop. (%)	1.8[g]	...
Annual growth of econ. active male pop. (%)	0.5[g]	...
Labour force in industry (%)	35.3[h]	30.4[g]
Labour force in agriculture (%)	4.7[h]	3.3[g]
Commercial energy production (1,000 Mt coal equiv)	12[f]	98[e]
Motor vehicles (per 1,000 population)	486	604[e]
Telephone lines (per 100 inhabitants)	43.8[i]	51.1[e]

Social indicators	1990/95
Growth rate of population (% per annum)	0.7
Age group 0-14 years (%)	17.6
Age group 60+ years (women and men, %)	11.7/7.8
Life expectancy at birth (women and men, years)	79/72
Infant mortality rate (per 1,000 births)	8
Total fertility rate (births per woman)	2
Urban population (%)	86
Urban population growth rate (% per annum)	1.1
Rural population growth rate (% per annum)	−1.7
Foreign-born (1985,%)	27.1
Refugees	2200
Government education expenditure (% of GDP)	4[j]
Primary-secondary gross enrolment ratio (f and m, per 100)	82/80[i]
Third-level students (per 100,000 population)	245[k]
Newspaper circulation (per 1,000 population)	383
Television receivers (per 1,000 population)	267
Intentional homicides (1986, per 100,000 population)	2[l]
Parliamentary seats (women and men, %)	13/87

Environment	1990/91
Threatened species	14
CO_2 emissions (10,000 Mt)	2814
Energy consumption (1,000 Mt coal equiv)	13757
Precipitation (mm)	740[i]
Average temperature (January and July, centigrade)	0.3/17.4[f]

a Capital city. b July 1994. c 1988. d 1992. e 1991. f 1980. g 1990.
h 1983. i 1987. j 1989. k 1985. l Source: DYB92.

Macau

Region	Eastern Asia
Location (longitude, latitude)	22°12'N 113°33'E[a]
Currency	pataca
Population (1995 est., in thousands)	444
Surface area (square kms)	16
Population density (pop. per square km)	31063
Sex ratio (females per 100 males)	94
Largest city (pop. in thousands)	Macau (457)[a]
Major language(s)	Chinese, Portuguese

Economic indicators	1985	1994
Consumer price index (1980=100)	102	182[b]
Tourist arrivals (in thousands)	2273[c]	3180[d]
Agricultural production index (1979-1981=100)	112[c]	73[d]
Food production index (1979-1981=100)	112[c]	73[d]
Motor vehicles (per 1,000 population)	70[g]	74[f]
Telephone lines (per 100 inhabitants)	13.2[h]	22.3[f]

Largest export industries		Major trading partners			1992
	(% of exports)		(% of exports)		(% of imports)
Textiles	80	United States	35	Hong Kong	33
Metal manufacture	8	Hong Kong	13	China	20
Other manufacturing	6	Germany	12	Japan	18

Social indicators	1990/95
Age group 0-14 years (%)	21.8
Age group 60+ years (women and men, %)	4.9/3.6[ci]
Life expectancy at birth (women and men, years)	80/75[j]
Infant mortality rate (per 1,000 births)	9
Urban population (%)	99
Urban population growth rate (% per annum)	2.9
Rural population growth rate (% per annum)	0.5
Foreign-born (1985,%)	39.0
Newspaper circulation (per 1,000 population)	518
Television receivers (per 1,000 population)	67
Intentional homicides (1986, per 100,000 population)	2[k]

Environment	1990/91
CO_2 emissions (10,000 Mt)	297
Energy consumption (1,000 Mt coal equiv)	1092
Precipitation (mm)	1846[h]
Average temperature (January and July, centigrade)	15.1/28.5[e]

a Capital city. b May 1994. c 1988. d 1992. e 1980. f 1991. g 1990.
h 1987. i Source DYB90. j 1985. k Source: DYB92.

Madagascar

Region Eastern Africa
Location (longitude, latitude) 15°91'S 47°53'E[a]
Currency Malagasy franc
Population (1995 est., in thousands) 14155
Surface area (square kms) 587041
Population density (pop. per square km) 20
Sex ratio (females per 100 males) 102
Largest city (pop. in thousands) Antananarivo (687)[a]
UN membership date 20-Sep-1960
Major language(s) Malagasy

Economic indicators	1985	1994
Exchange rate (US$)	635.79	3718.60
Consumer price index (1980=100)	249	1063
Balance of payments, current account (million US$)	−184	−136[b]
Tourist arrivals (in thousands)	35[c]	54[b]
GDP (million US$)	2345	2673[d]
GDP (per capita US$)	229	215[d]
Long-term rate of change in GDP (% per annum)	2.3	−6.6[d]
Economically active female population (%)	60.8[e]	53.4
Economically active male population (%)	90.1[e]	88.5
Annual growth of econ. active female pop. (%)	1.9[f]	...
Annual growth of econ. active male pop. (%)	2.3[f]	...
Agricultural production index (1979-1981=100)	116[c]	124[b]
Food production index (1979-1981=100)	117[c]	126[b]
Commercial energy production (1,000 Mt coal equiv)	18[e]	39[d]
Motor vehicles (per 1,000 population)	9	...
Telephone lines (per 100 inhabitants)	0.2[g]	0.3[d]

Social indicators	1990/95
Growth rate of population (% per annum)	3.3
Age group 0-14 years (%)	45.7
Age group 60+ years (women and men, %)	2.5/2.1
Life expectancy at birth (women and men, years)	57/54
Infant mortality rate (per 1,000 births)	110
Total fertility rate (births per woman)	7
Contraceptive use (% of currently married women)	17
Urban population (%)	27
Urban population growth rate (% per annum)	5.6
Rural population growth rate (% per annum)	2.4
Foreign-born (1985,%)	0.3
Government education expenditure (% of GDP)	2
Primary-secondary gross enrolment ratio (f and m, per 100)	52/55[h]
Third-level students (per 100,000 population)	298
Newspaper circulation (per 1,000 population)	4
Television receivers (per 1,000 population)	20
Intentional homicides (1986, per 100,000 population)	7

Environment	1990/91
Threatened species	127
Forested area (%)	26.5
CO_2 emissions (10,000 Mt)	293
Energy consumption (1,000 Mt coal equiv)	43
Precipitation (mm)	1270[g]
Average temperature (January and July, centigrade)	19.3/13.0[e]

a Capital city. b 1992. c 1988. d 1991. e 1980. f 1990. g 1987. h 1989.

Malawi

Region	Eastern Africa
Location (longitude, latitude)	13°58'S 33°42'E[a]
Currency	Malawi kwacha
Population (1995 est., in thousands)	11304
Surface area (square kms)	118484
Population density (pop. per square km)	72
Sex ratio (females per 100 males)	102
Largest city (pop. in thousands)	Lilongwe (310)[a]
UN membership date	01-Sep-1964
Major language(s)	Nyanja, English, Makua, Yao

Economic indicators	1985	1994
Exchange rate (US$)	1.68	7.39
Consumer price index (1980=100)	185	201[b]
Industrial production index (1980 = 100)	100	132[c]
Balance of payments, current account (million US$)	−125	...
Tourist arrivals (in thousands)	99[d]	135[c]
GDP (million US$)	1122	2191[e]
GDP (per capita US$)	153	219[e]
Long-term rate of change in GDP (% per annum)	7.8	0.0[e]
Gross fixed capital formation (% of GDP)	13.5	12.0[f]
Economically active female population (%)	62.4[g]	54.7
Economically active male population (%)	89.9[g]	88.2
Annual growth of econ. active female pop. (%)	2.1[f]	...
Annual growth of econ. active male pop. (%)	2.9[f]	...
Labour force in industry (%)	20.9[g]	21.5[f]
Labour force in agriculture (%)	49.2[g]	40.0[f]
Agricultural production index (1979-1981=100)	114[d]	98[c]
Food production index (1979-1981=100)	110[d]	86[c]
Commercial energy production (1,000 Mt coal equiv)	48[g]	90[e]
Motor vehicles (per 1,000 population)	4	5[f]
Telephone lines (per 100 inhabitants)	0.3[h]	0.3[f]

Social indicators	1990/95
Growth rate of population (% per annum)	3.3
Age group 0-14 years (%)	49.2
Age group 60+ years (women and men, %)	2.3/1.8
Life expectancy at birth (women and men, years)	45/44
Infant mortality rate (per 1,000 births)	142
Total fertility rate (births per woman)	8
Contraceptive use (% of currently married women)	13[d]
Urban population (%)	14
Urban population growth rate (% per annum)	6.0
Rural population growth rate (% per annum)	2.9
Foreign-born (1985,%)	3.9
Refugees	1058500
Government education expenditure (% of GDP)	2
Primary-secondary gross enrolment ratio (f and m, per 100)	47/58[i]
Third-level students (per 100,000 population)	61[d]
Newspaper circulation (per 1,000 population)	3
Intentional homicides (1986, per 100,000 population)	2
Parliamentary seats (women and men, %)	12/88

Environment	1990/91
Threatened species	16
Forested area (%)	30.6
CO_2 emissions (10,000 Mt)	172
Energy consumption (1,000 Mt coal equiv)	37
Precipitation (mm)	849[h]
Average temperature (January and July, centigrade)	21.2/15.0[g]

a Capital city. b June 1994. c 1992. d 1988. e 1991. f 1990. g 1980.
h 1987. i 1989.

Malaysia

Region	South-eastern Asia
Location (longitude, latitude)	3°14'N 101°71'E
Currency	ringgit
Population (1995 est., in thousands)	20125
Surface area (square kms)	329749
Population density (pop. per square km)	56
Sex ratio (females per 100 males)	98
Largest city (pop. in thousands)	Kuala Lumpur (1684)
UN membership date	17-Sep-1957
Major language(s)	Malay, Chinese, Tamil

Economic indicators	1985	1994
Exchange rate (US$)	2.43	2.57
Consumer price index (1980=100)	125	117
Industrial production index (1980 = 100)	138	362[a]
Balance of payments, current account (million US$)	−613	−2103[b]
Tourist arrivals (in thousands)	3624[c]	6016[d]
GDP (million US$)	31200	47104[e]
GDP (per capita US$)	1990	2568[e]
Long-term rate of change in GDP (% per annum)	−1.1	8.7[e]
Gross fixed capital formation (% of GDP)	29.8	35.5[e]
Economically active female population (%)	43.0[f]	45.3
Economically active male population (%)	82.7[f]	82.7
Annual growth of econ. active female pop. (%)	4.2[g]	...
Annual growth of econ. active male pop. (%)	3.1[g]	...
Labour force in industry (%)	23.8[f]	27.5[g]
Labour force in agriculture (%)	37.2[f]	26.0[g]
Agricultural production index (1979-1981=100)	164[c]	205[d]
Food production index (1979-1981=100)	187[c]	255[d]
Commercial energy production (1,000 Mt coal equiv)	19474[f]	66531[e]
Motor vehicles (per 1,000 population)	108	135[e]
Telephone lines (per 100 inhabitants)	6.8[h]	9.9[e]

Social indicators	1990/95
Growth rate of population (% per annum)	2.4
Age group 0-14 years (%)	37.9
Age group 60+ years (women and men, %)	3.2/2.7
Life expectancy at birth (women and men, years)	73/69
Infant mortality rate (per 1,000 births)	14
Total fertility rate (births per woman)	4
Contraceptive use (% of currently married women)	48
Urban population (%)	47
Urban population growth rate (% per annum)	4.2
Rural population growth rate (% per annum)	0.8
Foreign-born (1985,%)	4.7
Refugees	10300
Government education expenditure (% of GDP)	5
Primary-secondary gross enrolment ratio (f and m, per 100)	76/74
Third-level students (per 100,000 population)	671
Newspaper circulation (per 1,000 population)	140
Television receivers (per 1,000 population)	149
Intentional homicides (1986, per 100,000 population)	2
Parliamentary seats (women and men, %)	5/95

Environment	1990/91
Threatened species	110
Forested area (%)	58.6
CO_2 emissions (10,000 Mt)	16702
Energy consumption (1,000 Mt coal equiv)	1534
Precipitation (mm)	2499[h]
Average temperature (January and July, centigrade)	26.8/27.1[f]

a July 1994. b 1993. c 1988. d 1992. e 1991. f 1980. g 1990. h 1987.

Maldives

Region	Southern Asia
Location (longitude, latitude)	4°10'N 73°30'E[a]
Currency	rufiyaa
Population (1995 est., in thousands)	248
Surface area (square kms)	298
Population density (pop. per square km)	748
Sex ratio (females per 100 males)	92
Largest city (pop. in thousands)	Male (63)[a]
UN membership date	21-Sep-1965
Major language(s)	Maldivian

Economic indicators	1985	1994
Exchange rate (US$)	7.13	11.86
Balance of payments, current account (million US$)	−6	−33[b]
Tourist arrivals (in thousands)	156[c]	236[b]
GDP (million US$)	84	152[d]
GDP (per capita US$)	459	689[d]
Long-term rate of change in GDP (% per annum)	13.8	4.7[d]
Economically active female population (%)	27.1[e]	24.7
Economically active male population (%)	83.2[e]	82.4
Agricultural production index (1979-1981=100)	121[c]	128[b]
Food production index (1979-1981=100)	121[c]	128[b]
Telephone lines (per 100 inhabitants)	2.9[f]	3.5[d]

Social indicators	1990/95
Growth rate of population (% per annum)	3.0
Age group 0-14 years (%)	44.0
Age group 60+ years (women and men, %)	2.8/2.8
Life expectancy at birth (women and men, years)	62/65
Infant mortality rate (per 1,000 births)	55
Total fertility rate (births per woman)	6
Urban population (%)	33
Urban population growth rate (% per annum)	5.5
Rural population growth rate (% per annum)	1.9
Foreign-born (1985,%)	1.4
Newspaper circulation (per 1,000 population)	8
Television receivers (per 1,000 population)	25
Parliamentary seats (women and men, %)	4/96

Environment	1990/91
Threatened species	5
Forested area (%)	3.3
CO_2 emissions (10,000 Mt)	26
Energy consumption (1,000 Mt coal equiv)	205

a Capital city. b 1992. c 1988. d 1991. e 1980. f 1990.

Mali

Region	Western Africa
Location (longitude, latitude)	12°65'N 8°01'W[a]
Currency	CFA franc
Population (1995 est., in thousands)	10797
Surface area (square kms)	1240192
Population density (pop. per square km)	8
Sex ratio (females per 100 males)	103
Largest city (pop. in thousands)	Bamako (738)[a]
UN membership date	28-Sep-1960
Major language(s)	English, Bambara, Ful, Senufo, Soninke, Malinke, Songhai

Economic indicators	1985	1994
Exchange rate (US$)	378.05	528.15
Consumer price index (1980=100)	101[b]	119[c]
Balance of payments, current account (million US$)	−129	−91[d]
Tourist arrivals (in thousands)	36[e]	38[d]
GDP (million US$)	1158	2451[f]
GDP (per capita US$)	146	258[f]
Long-term rate of change in GDP (% per annum)	8.5	−0.2[f]
Gross fixed capital formation (% of GDP)	19.2	20.0[f]
Economically active female population (%)	16.7[g]	15.1
Economically active male population (%)	91.2[g]	90.3
Annual growth of econ. active female pop. (%)	1.8[b]	...
Annual growth of econ. active male pop. (%)	2.2[b]	...
Agricultural production index (1979-1981=100)	128[e]	137[d]
Food production index (1979-1981=100)	127[e]	132[d]
Commercial energy production (1,000 Mt coal equiv)	10[g]	23[f]
Motor vehicles (per 1,000 population)	4	...
Telephone lines (per 100 inhabitants)	0.1[h]	0.1[f]

Social indicators	1990/95
Growth rate of population (% per annum)	3.2
Age group 0-14 years (%)	47.4
Age group 60+ years (women and men, %)	2.3/1.8
Life expectancy at birth (women and men, years)	48/44
Infant mortality rate (per 1,000 births)	159
Total fertility rate (births per woman)	7
Contraceptive use (% of currently married women)	5[h]
Urban population (%)	27
Urban population growth rate (% per annum)	5.7
Rural population growth rate (% per annum)	2.3
Foreign-born (1985,%)	1.4
Refugees	13100
Government education expenditure (% of GDP)	3[h]
Primary-secondary gross enrolment ratio (f and m, per 100)	11/20[i]
Third-level students (per 100,000 population)	73
Newspaper circulation (per 1,000 population)	1
Television receivers (per 1,000 population)	1
Parliamentary seats (women and men, %)	2/98

Environment	1990/91
Threatened species	20
Forested area (%)	5.6
CO_2 emissions (10,000 Mt)	119
Energy consumption (1,000 Mt coal equiv)	24
Precipitation (mm)	1099[h]
Average temperature (January and July, centigrade)	25.5/26.9[g]

a Capital city. b 1990. c March 1994. d 1992. e 1988. f 1991. g 1980. h 1987. i 1989.

Poland

Region	Eastern Europe
Location (longitude, latitude)	50°16'N 19°00'E
Currency	zloty
Population (1995 est., in thousands)	38736
Surface area (square kms)	323250
Population density (pop. per square km)	118
Sex ratio (females per 100 males)	105
Largest city (pop. in thousands)	Katowice (3449)
UN membership date	24-Oct-1945
Major language(s)	Polish

Economic indicators	1985	1994
Exchange rate (US$)	151.00	22700.00
Consumer price index (1980=100)	393	469
Industrial production index (1980 = 100)	99	84[a]
Unemployment (%)	...	16.5
Balance of payments, current account (million US$)	−982	−3104[b]
Tourist arrivals (in thousands)	2495[c]	4000[b]
GDP (million US$)	69226	77943[d]
GDP (per capita US$)	1861	2035[d]
Long-term rate of change in GDP (% per annum)	3.6	−7.6[d]
Gross fixed capital formation (% of GDP)	21.2	18.8[d]
Economically active female population (%)	58.6[e]	57.2
Economically active male population (%)	76.7[e]	74.7
Annual growth of econ. active female pop. (%)	0.7[f]	...
Annual growth of econ. active male pop. (%)	0.6[f]	
Labour force in industry (%)	40.0[g]	35.4[d]
Labour force in agriculture (%)	29.9[g]	26.7[d]
Agricultural production index (1979-1981=100)	111[c]	102[b]
Food production index (1979-1981=100)	112[c]	105[b]
Commercial energy production (1,000 Mt coal equiv)	173830[e]	132130[d]
Motor vehicles (per 1,000 population)	122	193[d]
Telephone lines (per 100 inhabitants)	7.4[h]	9.3[d]

Largest export industries	Major trading partners			1992
(% of exports)	(% of exports)		(% of imports)	
...	Germany	31	Germany	24
...	Netherlands	6	Italy	7
...	Italy	6	United Kingdom	7

Social indicators	1990/95
Growth rate of population (% per annum)	0.3
Age group 0-14 years (%)	23.5
Age group 60+ years (women and men, %)	9.4/6.3
Life expectancy at birth (women and men, years)	76/67
Infant mortality rate (per 1,000 births)	15
Total fertility rate (births per woman)	2
Urban population (%)	64
Urban population growth rate (% per annum)	1.0
Rural population growth rate (% per annum)	−0.8
Foreign-born (1985, %)	4.0
Refugees	2700
Government education expenditure (% of GDP)	4
Primary-secondary gross enrolment ratio (f and m, per 100)	94/93
Third-level students (per 100,000 population)	1418
Newspaper circulation (per 1,000 population)	128
Television receivers (per 1,000 population)	295
Intentional homicides (1986, per 100,000 population)	1
Parliamentary seats (women and men, %)	10/90

Environment	1990/91
Threatened species	55
Forested area (%)	28.0
CO_2 emissions (10,000 Mt)	84106
Energy consumption (1,000 Mt coal equiv)	3167
Precipitation (mm)	613[h]
Average temperature (January and July, centigrade)	−2.4/18.7[e]

a June 1994. b 1992. c 1988. d 1991. e 1980. f 1990. g 1981. h 1987.

Portugal

Region Southern Europe
Location (longitude, latitude) 38°72'N 9°13'W
Currency Portuguese escudo
Population (1995 est., in thousands) 9884
Surface area (square kms) 92389
Population density (pop. per square km) 115
Sex ratio (females per 100 males) 107
Largest city (pop. in thousands) Lisbon (1580)
UN membership date 14-Dec-1955
Major language(s) Portuguese

Economic indicators	1985	1994
Exchange rate (US$)	157.49	157.86
Consumer price index (1980=100)	284	123
Industrial production index (1980 = 100)	119	153[a]
Unemployment (%)	...	6.7[b]
Balance of payments, current account (million US$)	380	−184[c]
Tourist arrivals (in thousands)	6624[d]	8884[c]
GDP (million US$)	20682	68614[e]
GDP (per capita US$)	2088	6955[e]
Long-term rate of change in GDP (% per annum)	2.8	2.2[e]
Gross fixed capital formation (% of GDP)	21.8	26.4[f]
Economically active female population (%)	38.5[g]	39.3
Economically active male population (%)	81.0[g]	77.6
Annual growth of econ. active female pop. (%)	3.9[f]	...
Annual growth of econ. active male pop. (%)	0.9[f]	...
Labour force in industry (%)	36.4[g]	33.0[c]
Labour force in agriculture (%)	27.3[g]	11.5[c]
Agricultural production index (1979-1981=100)	95[d]	119[c]
Food production index (1979-1981=100)	95[d]	119[c]
Commercial energy production (1,000 Mt coal equiv)	1170[g]	1292[c]
Motor vehicles (per 1,000 population)	239	370[e]
Telephone lines (per 100 inhabitants)	16.7[h]	27.3[e]

Largest export industries		Major trading partners			1992
	(% of exports)		(% of exports)		(% of imports)
Textiles	39	Germany	19	Spain	17
Metal manufacture	26	Spain	15	Germany	15
Chemicals	9	France[i]	14	France[i]	13

Social indicators	1990/95
Growth rate of population (% per annum)	0.3
Age group 0-14 years (%)	18.5
Age group 60+ years (women and men, %)	11.3/8.1
Life expectancy at birth (women and men, years)	78/71
Infant mortality rate (per 1,000 births)	12
Total fertility rate (births per woman)	1
Urban population (%)	36
Urban population growth rate (% per annum)	1.6
Rural population growth rate (% per annum)	−0.8
Foreign-born (1985,%)	0.8
Refugees	1800
Government education expenditure (% of GDP)	5
Primary-secondary gross enrolment ratio (f and m, per 100)	86/88[j]
Third-level students (per 100,000 population)	1882
Newspaper circulation (per 1,000 population)	39
Television receivers (per 1,000 population)	187
Intentional homicides (1986, per 100,000 population)	5
Parliamentary seats (women and men, %)	9/91

Environment	1990/91
Threatened species	118
Forested area (%)	32.1
CO_2 emissions (10,000 Mt)	11406
Energy consumption (1,000 Mt coal equiv)	1960
Precipitation (mm)	708[h]
Average temperature (January and July, centigrade)	10.8/22.2[g]

a March 1994. b Labour force sample surveys. c 1992. d 1988. e 1991.
f 1990. g 1980. h 1987. i Includes Monaco. j 1989.

Puerto Rico

Region	Caribbean
Location (longitude, latitude)	18°47'N 66°11'W
Currency	US dollar
Population (1995 est., in thousands)	3691
Surface area (square kms)	8897
Population density (pop. per square km)	405
Sex ratio (females per 100 males)	105
Largest city (pop. in thousands)	San Juan (1383)
Major language(s)	English, Spanish

Economic indicators	1985	1994
Consumer price index (1980=100)	117	152[a]
Unemployment (%)	...	16.0[bc]
Tourist arrivals (in thousands)	2281[d]	2640[e]
GDP (million US$)	21969	33969[c]
GDP (per capita US$)	6505	9537[c]
Long-term rate of change in GDP (% per annum)	8.2	3.1[c]
Gross fixed capital formation (% of GDP)	10.6	15.2[c]
Economically active female population (%)	25.5[f]	25.6
Economically active male population (%)	68.0[f]	68.4
Annual growth of econ. active female pop. (%)	2.7[g]	...
Annual growth of econ. active male pop. (%)	2.1[g]	...
Labour force in industry (%)	26.2[f]	24.0[e]
Labour force in agriculture (%)	5.3[f]	3.3[e]
Agricultural production index (1979-1981=100)	101[d]	103[e]
Food production index (1979-1981=100)	100[d]	103[e]
Commercial energy production (1,000 Mt coal equiv)	17[f]	36[c]
Motor vehicles (per 1,000 population)	388	443[c]
Telephone lines (per 100 inhabitants)	20.4[h]	28.9[c]

Social indicators	1990/95
Growth rate of population (% per annum)	0.9
Age group 0-14 years (%)	26.3
Age group 60+ years (women and men, %)	6.8/5.3
Life expectancy at birth (women and men, years)	78/72
Infant mortality rate (per 1,000 births)	13
Total fertility rate (births per woman)	2
Urban population (%)	77
Urban population growth rate (% per annum)	1.6
Rural population growth rate (% per annum)	−1.3
Foreign-born (1985,%)	10.4
Third-level students (per 100,000 population)	4091[f]
Newspaper circulation (per 1,000 population)	129
Television receivers (per 1,000 population)	264
Intentional homicides (1986, per 100,000 population)	23[i]

Environment	1990/91
Threatened species	21
Forested area (%)	19.9
CO_2 emissions (10,000 Mt)	3277
Energy consumption (1,000 Mt coal equiv)	2030
Precipitation (mm)	1631[h]
Average temperature (January and July, centigrade)	23.6/26.9[f]

a April 1994. b Labour force sample surveys. c 1991. d 1988. e 1992. f 1980. g 1990. h 1987. i Source: DYB92.

Qatar

Region	Western Asia
Location (longitude, latitude)	25°17'N 51°32'E[a]
Currency	Qatari riyal
Population (1995 est., in thousands)	490
Surface area (square kms)	11000
Population density (pop. per square km)	35
Sex ratio (females per 100 males)	55
Largest city (pop. in thousands)	Doha (293)[a]
UN membership date	21-Sep-1971
Major language(s)	Arabic, Persian

Economic indicators	1985	1994
Exchange rate (US$)	3.64	3.64
Consumer price index (1980=100)	112	...
Tourist arrivals (in thousands)	113[b]	141[c]
GDP (million US$)	6153	6673[d]
GDP (per capita US$)	17188	15165[d]
Long-term rate of change in GDP (% per annum)	−3.9	−1.5[d]
Gross fixed capital formation (% of GDP)	17.7	
Economically active female population (%)	14.0[e]	18.7
Economically active male population (%)	93.7[e]	94.7
Annual growth of econ. active female pop. (%)	13.7[f]	...
Annual growth of econ. active male pop. (%)	6.6[f]	
Commercial energy production (1,000 Mt coal equiv)	38746[e]	40841[d]
Motor vehicles (per 1,000 population)	317	357[f]
Telephone lines (per 100 inhabitants)	19.8[g]	21.9[d]

Social indicators	1990/95
Growth rate of population (% per annum)	2.8
Age group 0-14 years (%)	29.6
Age group 60+ years (women and men, %)	0.8/2.2
Life expectancy at birth (women and men, years)	73/68
Infant mortality rate (per 1,000 births)	26
Total fertility rate (births per woman)	4
Contraceptive use (% of currently married women)	32[g]
Urban population (%)	91
Urban population growth rate (% per annum)	3.1
Rural population growth rate (% per annum)	−0.5
Foreign-born (1985,%)	31.9
Primary-secondary gross enrolment ratio (f and m, per 100)	95/93
Third-level students (per 100,000 population)	1848[h]
Newspaper circulation (per 1,000 population)	187
Television receivers (per 1,000 population)	450
Intentional homicides (1986, per 100,000 population)	2

Environment	1990/91
Threatened species	7
CO_2 emissions (10,000 Mt)	5362
Energy consumption (1,000 Mt coal equiv)	29161

a Capital city. b 1988. c 1992. d 1991. e 1980. f 1990. g 1987. h 1989.

Republic of Moldova

Region Eastern Europe
Location (longitude, latitude) 47°00'N 28°50'E[a]
Currency rouble
Surface area (square kms) 33700
Population density (pop. per square km) 129
Largest city (pop. in thousands) Kishinev (685)[a]
UN membership date 02-Mar-1992
Major language(s) Romanian, Ukranian, Russian

Economic indicators	1985	1994
Exchange rate (US$)	...	4.01
Consumer price index (1980=100)	...	138947[b]
GDP (million US$)	10071	13188[c]
GDP (per capita US$)	2389	3021[c]
Long-term rate of change in GDP (% per annum)	−9.8	−12.6[c]
Economically active female population (%)	66.8[d]	65.4
Economically active male population (%)	80.5[d]	80.8
Labour force in industry (%)	25.9[e]	25.8[f]
Labour force in agriculture (%)	39.9[e]	40.0[f]
Motor vehicles (per 1,000 population)	39	53[c]
Telephone lines (per 100 inhabitants)	8.7[g]	11.3[c]

Social indicators	1990/95
Contraceptive use (% of currently married women)	22

Environment	1990/91
Threatened species	31

a Capital city. b July 1994. c 1991. d 1980. e 1981. f 1992. g 1987.

Réunion

Region	Eastern Africa
Location (longitude, latitude)	20°52'S 55°28'E[a]
Currency	French franc
Population (1995 est., in thousands)	653
Surface area (square kms)	2510
Population density (pop. per square km)	242
Sex ratio (females per 100 males)	104
Largest city (pop. in thousands)	Saint-Denis (123)[a]
Major language(s)	French

Economic indicators	1985	1994
Consumer price index (1980=100)	158	207
Tourist arrivals (in thousands)	175[b]	217[c]
GDP (million US$)	1898	5075[d]
GDP (per capita US$)	3451	8265[d]
Long-term rate of change in GDP (% per annum)	3.5	3.5[d]
Gross fixed capital formation (% of GDP)	21.9	...
Economically active female population (%)	33.0[e]	37.3
Economically active male population (%)	73.8[e]	75.7
Annual growth of econ. active female pop. (%)	5.8[f]	...
Annual growth of econ. active male pop. (%)	2.3[f]	...
Agricultural production index (1979-1981=100)	108[b]	109[c]
Food production index (1979-1981=100)	108[b]	109[c]
Commercial energy production (1,000 Mt coal equiv)	37[e]	75[d]
Motor vehicles (per 1,000 population)	267	258[d]
Telephone lines (per 100 inhabitants)	21.2[g]	28.5[d]

Largest export industries	Major trading partners		1992
(% of exports)	(% of exports)		(% of imports)
Food, beverages, tobacco 73	France[d] 76	France[d]	64
Metal manufacture 14	Japan 5	Bahrain	4
Agriculture 7	Comoros 4	Italy	3

Social indicators	1990/95
Growth rate of population (% per annum)	1.6
Age group 0-14 years (%)	29.4
Age group 60+ years (women and men, %)	5.5/3.5
Life expectancy at birth (women and men, years)	78/69
Infant mortality rate (per 1,000 births)	7
Total fertility rate (births per woman)	2
Contraceptive use (% of currently married women)	67
Urban population (%)	68
Urban population growth rate (% per annum)	2.8
Rural population growth rate (% per annum)	−0.7
Foreign-born (1985,%)	7.9
Newspaper circulation (per 1,000 population)	108
Television receivers (per 1,000 population)	163

Environment	1990/91
Threatened species	26
Forested area (%)	35.1
CO_2 emissions (10,000 Mt)	302
Energy consumption (1,000 Mt coal equiv)	990
Precipitation (mm)	1526[g]
Average temperature (January and July, centigrade)	26.0/20.7[e]

a Capital city. b 1988. c 1992. d 1991. e 1980. f 1990. g 1987.

Romania

Region	Eastern Europe
Location (longitude, latitude)	44°44'N 26°09'E
Currency	lei
Population (1995 est., in thousands)	23505
Surface area (square kms)	237500
Population density (pop. per square km)	98
Sex ratio (females per 100 males)	102
Largest city (pop. in thousands)	Bucharest (2201)
UN membership date	14-Dec-1955
Major language(s)	Romanian, Hungarian

Economic indicators	1985	1994
Exchange rate (US$)	15.73	460.00
Consumer price index (1980=100)	91	7468[a]
Industrial production index (1980 = 100)	120	69[b]
Balance of payments, current account (million US$)	1381	−1506[b]
Tourist arrivals (in thousands)	5514[c]	6280[b]
GDP (million US$)	47687	27617[d]
GDP (per capita US$)	2098	1187[d]
Long-term rate of change in GDP (% per annum)	−0.1	−13.7[d]
Gross fixed capital formation (% of GDP)	30.1	14.9[d]
Economically active female population (%)	58.8[e]	54.0
Economically active male population (%)	75.2[e]	70.5
Annual growth of econ. active female pop. (%)	0.7[f]	...
Annual growth of econ. active male pop. (%)	0.1[f]	...
Labour force in industry (%)	43.8[e]	39.9[d]
Labour force in agriculture (%)	29.8[e]	29.8[d]
Agricultural production index (1979-1981=100)	105[c]	67[b]
Food production index (1979-1981=100)	104[c]	67[b]
Commercial energy production (1,000 Mt coal equiv)	84192[e]	49406[d]
Telephone lines (per 100 inhabitants)	9.5[g]	10.5[d]

Social indicators	1990/95
Growth rate of population (% per annum)	0.3
Age group 0-14 years (%)	22.2
Age group 60+ years (women and men, %)	9.3/7.3
Life expectancy at birth (women and men, years)	73/67
Infant mortality rate (per 1,000 births)	23
Total fertility rate (births per woman)	2
Urban population (%)	56
Urban population growth rate (% per annum)	1.2
Rural population growth rate (% per annum)	−0.9
Foreign-born (1985, %)	0.7
Refugees	500
Government education expenditure (% of GDP)	3
Primary-secondary gross enrolment ratio (f and m, per 100)	94/88
Third-level students (per 100,000 population)	694[c]
Newspaper circulation (per 1,000 population)	158[e]
Television receivers (per 1,000 population)	196
Intentional homicides (1986, per 100,000 population)	5[h]
Parliamentary seats (women and men, %)	4/96

Environment	1990/91
Threatened species	55
Forested area (%)	26.9
CO_2 emissions (10,000 Mt)	37671
Energy consumption (1,000 Mt coal equiv)	3025
Precipitation (mm)	578[g]
Average temperature (January and July, centigrade)	−2.7/23.3[e]

a August 1994. b 1992. c 1988. d 1991. e 1980. f 1990. g 1987.
h Source: DYB92.

Russian Federation

Location (longitude, latitude) 55°76'N 37°62'E
Currency rouble
Population (1995 est., in thousands) 288562[a]
Surface area (square kms) 17075400
Sex ratio (females per 100 males) 110[a]
Largest city (pop. in thousands) Moscow (9048)
UN membership date 24-Oct-1945
Major language(s) Russian

Economic indicators	1985	1994
Exchange rate (US$)	0.77[a]	2144.00[a]
Consumer price index (1980=100)	105[a]	118[ab]
Industrial production index (1980 = 100)	119[a]	124[c]
Tourist arrivals (in thousands)[a]	6007[d]	6900[e]
GDP (million US$)	584574	745470[c]
GDP (per capita US$)	4079	5012[c]
Long-term rate of change in GDP (% per annum)[a]	1.8	−0.9[c]
Economically active female population (%)	56.8[f]	54.8
Economically active male population (%)	79.2[f]	77.9
Annual growth of econ. active female pop. (%)	0.9[ab]	...
Annual growth of econ. active male pop. (%)	1.4[ab]	...
Labour force in industry (%)	38.3[af]	38.3[c]
Labour force in agriculture (%)	20.1[af]	14.2[c]
Agricultural production index (1979-1981=100)[a]	117[d]	99[e]
Food production index (1979-1981=100)[a]	118[d]	101[e]
Commercial energy production (1,000 Mt coal equiv)	1995707[af]	2217666[ac]
Motor vehicles (per 1,000 population)[a]	48	...
Telephone lines (per 100 inhabitants)	11.5[g]	15.0[c]

Social indicators	1990/95
Growth rate of population (% per annum)[a]	0.5
Age group 0-14 years (%)[a]	25.1
Age group 60+ years (women and men, %)[a]	9.8/5.3
Life expectancy at birth (women and men, years)[a]	75/66
Infant mortality rate (per 1,000 births)[a]	21
Total fertility rate (births per woman)[a]	2
Contraceptive use (% of currently married women)[a]	32
Urban population (%)[a]	68
Urban population growth rate (% per annum)[a]	1.1
Rural population growth rate (% per annum)[a]	−0.7
Foreign-born (1985, %)[a]	0.6
Refugees	17100
Government education expenditure (% of GDP)[a]	7
Primary-secondary gross enrolment ratio (f and m, per 100)[a]	93/87
Third-level students (per 100,000 population)[a]	1820
Newspaper circulation (per 1,000 population)[a]	493[h]
Television receivers (per 1,000 population)[a]	305[i]
Intentional homicides (1986, per 100,000 population)	5[a]
Parliamentary seats (women and men, %)[a]	35/65[g]

Environment	1990/91
Threatened species	112
Forested area (%)[a]	42.3
CO_2 emissions (10,000 Mt)[a]	977396
Energy consumption (1,000 Mt coal equiv)[a]	29607
Precipitation (mm)	575[a]
Average temperature (January and July, centigrade)[a]	−9.9/19.0[f]

a Data refer to former USSR. b 1990. c 1991. d 1988. e 1992. f 1980.
g 1987. h 1989. i 1985.

Rwanda

Region	Eastern Africa
Location (longitude, latitude)	1°58'S 30°08'E[a]
Currency	Rwanda franc
Population (1995 est., in thousands)	8330
Surface area (square kms)	26338
Population density (pop. per square km)	284
Sex ratio (females per 100 males)	102
Largest city (pop. in thousands)	Kigali (219)[a]
UN membership date	18-Sep-1962
Major language(s)	Kinyarwanda, French

Economic indicators	1985	1994
Exchange rate (US$)	93.49	146.37
Consumer price index (1980=100)	129	216[b]
Balance of payments, current account (million US$)	−64	−85[c]
Tourist arrivals (in thousands)	36[d]	5[c]
GDP (million US$)	1715	1701[e]
GDP (per capita US$)	288	234[e]
Long-term rate of change in GDP (% per annum)	4.4	1.1[e]
Gross fixed capital formation (% of GDP)	15.6	11.9[e]
Economically active female population (%)	84.7[f]	76.8
Economically active male population (%)	93.9[f]	92.8
Annual growth of econ. active female pop. (%)	2.7[g]	...
Annual growth of econ. active male pop. (%)	3.2[g]	...
Agricultural production index (1979-1981=100)	107[d]	115[c]
Food production index (1979-1981=100)	102[d]	113[c]
Commercial energy production (1,000 Mt coal equiv)	15[f]	22[e]
Motor vehicles (per 1,000 population)	2	2[g]
Telephone lines (per 100 inhabitants)	0.1[h]	0.2[e]

Social indicators	1990/95
Growth rate of population (% per annum)	3.4
Age group 0-14 years (%)	49.8
Age group 60+ years (women and men, %)	2.0/1.7
Life expectancy at birth (women and men, years)	48/45
Infant mortality rate (per 1,000 births)	110
Total fertility rate (births per woman)	8
Contraceptive use (% of currently married women)	21
Urban population (%)	6
Urban population growth rate (% per annum)	5.0
Rural population growth rate (% per annum)	3.3
Foreign-born (1985,%)	2.0
Refugees	25200
Government education expenditure (% of GDP)	4[i]
Primary-secondary gross enrolment ratio (f and m, per 100)	46/47[i]
Third-level students (per 100,000 population)	48[i]
Newspaper circulation (per 1,000 population)	0
Parliamentary seats (women and men, %)	17/83

Environment	1990/91
Threatened species	28
Forested area (%)	21.0
CO_2 emissions (10,000 Mt)	119
Energy consumption (1,000 Mt coal equiv)	30
Precipitation (mm)	986[h]

a Capital city. b December 1993. c 1992. d 1988. e 1991. f 1980. g 1990. h 1987. i 1989.

Saint Kitts and Nevis

Region Caribbean
Location (longitude, latitude) 16°10'N 61°40'W[a]
Currency East Caribbean dollar
Population (1995 est., in thousands) 44
Surface area (square kms) 261
Population density (pop. per square km) 169
Sex ratio (females per 100 males) 95
Largest city (pop. in thousands) Basse-Terre (21)[a]
UN membership date 23-Sep-1983
Major language(s) English

Economic indicators	1985	1994
Exchange rate (US$)	2.70[b]	2.70[b]
Consumer price index (1980=100)	126	150
Balance of payments, current account (million US$)	-7	-21[c]
Tourist arrivals (in thousands)	70[d]	90[c]
GDP (million US$)	67	121[e]
GDP (per capita US$)	1552	2880[e]
Long-term rate of change in GDP (% per annum)	5.6	6.8[e]
Motor vehicles (per 1,000 population)	92	133[e]

Social indicators	1990/95
Growth rate of population (% per annum)	0.3
Age group 0-14 years (%)	32.1
Age group 60+ years (women and men, %)	7.1/5.2[dg]
Life expectancy at birth (women and men, years)	71/66[h]
Infant mortality rate (per 1,000 births)	24
Urban population (%)	53
Urban population growth rate (% per annum)	1.2
Rural population growth rate (% per annum)	-1.8
Foreign-born (1985,%)	5.9
Government education expenditure (% of GDP)	3[i]
Television receivers (per 1,000 population)	205
Intentional homicides (1986, per 100,000 population)	2
Parliamentary seats (women and men, %)	7/93

Environment	1990/91
Threatened species	5
Forested area (%)	16.7
CO_2 emissions (10,000 Mt)	20
Energy consumption (1,000 Mt coal equiv)	833

a Capital city. b Fixed rate. c 1992. d 1988. e 1991. f 1980. g Source DYB90. h 1985. i 1989.

Saint Lucia

Region	Caribbean
Location (longitude, latitude)	14°01'N 61°00'W[a]
Currency	East Caribbean dollar
Population (1995 est., in thousands)	148
Surface area (square kms)	622
Population density (pop. per square km)	246
Sex ratio (females per 100 males)	106
Largest city (pop. in thousands)	Castries (59)[a]
UN membership date	18-Sep-1979
Major language(s)	English, Creole French

Economic indicators	1985	1994
Exchange rate (US$)	2.70[b]	2.70[b]
Consumer price index (1980=100)	100	140[c]
Balance of payments, current account (million US$)	−13	−55[d]
Tourist arrivals (in thousands)	133[e]	177[d]
GDP (million US$)	167	275[f]
GDP (per capita US$)	1343	2037[f]
Long-term rate of change in GDP (% per annum)	6.0	1.0[f]
Agricultural production index (1979-1981=100)	182[e]	180[d]
Food production index (1979-1981=100)	182[e]	180[d]
Motor vehicles (per 1,000 population)	65	58[f]
Telephone lines (per 100 inhabitants)	12.8[h]	12.7[f]

Largest export industries	Major trading partners		1992
(% of exports)	(% of exports)		(% of imports)
...	United Kingdom	51	United States 34
...	United States	21	United Kingdom 14
...	Italy	6	Trinidad Tbg. 9

Social indicators	1990/95
Growth rate of population (% per annum)	1.2
Age group 0-14 years (%)	44.4
Age group 60+ years (women and men, %)	4.7/3.3[i][j]
Life expectancy at birth (women and men, years)	75/68[k]
Infant mortality rate (per 1,000 births)	18
Contraceptive use (% of currently married women)	47[e]
Urban population (%)	46
Urban population growth rate (% per annum)	2.3
Rural population growth rate (% per annum)	0.6
Foreign-born (1985, %)	3.1
Government education expenditure (% of GDP)	5[e]
Television receivers (per 1,000 population)	189
Intentional homicides (1986, per 100,000 population)	3
Parliamentary seats (women and men, %)	0/100

Environment	1990/91
Threatened species	9
Forested area (%)	12.9
CO_2 emissions (10,000 Mt)	44
Energy consumption (1,000 Mt coal equiv)	578

a Capital city. b Fixed rate. c July 1994. d 1992. e 1988. f 1991. g 1980. h 1990. i Source DYB90. j 1989. k 1985.

Saint Vincent/Grenadines

Region Caribbean
Location (longitude, latitude) 13°09'N 61°14'W[a]
Currency East Caribbean dollar
Population (1995 est., in thousands) 98
Surface area (square kms) 388
Population density (pop. per square km) 302
Sex ratio (females per 100 males) 106
Largest city (pop. in thousands) Kingstown (31)[a]
UN membership date 16-Sep-1980
Major language(s) English

Economic indicators	1985	1994
Exchange rate (US$)	2.70[b]	2.70[b]
Consumer price index (1980=100)	114[c]	130[d]
Balance of payments, current account (million US$)	4	−17[e]
Tourist arrivals (in thousands)	47[f]	53[e]
GDP (million US$)	113	178[g]
GDP (per capita US$)	1106	1648[g]
Long-term rate of change in GDP (% per annum)	4.4	2.6[g]
Gross fixed capital formation (% of GDP)	25.0	30.3[c]
Agricultural production index (1979-1981=100)	174[f]	168[e]
Food production index (1979-1981=100)	174[f]	168[e]
Commercial energy production (1,000 Mt coal equiv)	2[h]	5[g]
Motor vehicles (per 1,000 population)	63	70[g]
Telephone lines (per 100 inhabitants)	6.7[i]	13.9[g]

Social indicators	1990/95
Age group 0-14 years (%)	43.7[h]
Age group 60+ years (women and men, %)	4.8/3.4[h]
Infant mortality rate (per 1,000 births)	22
Contraceptive use (% of currently married women)	58[f]
Urban population (%)	22
Urban population growth rate (% per annum)	2.6
Rural population growth rate (% per annum)	0.4
Foreign-born (1985,%)	2.5
Government education expenditure (% of GDP)	5
Television receivers (per 1,000 population)	144
Intentional homicides (1986, per 100,000 population)	17
Parliamentary seats (women and men, %)	10/90

Environment	1990/91
Threatened species	5
Forested area (%)	35.9
CO_2 emissions (10,000 Mt)	21
Energy consumption (1,000 Mt coal equiv)	389

a Capital city. b Fixed rate. c 1990. d June 1994. e 1992. f 1988.
g 1991. h 1980. i 1987.

Samoa

Region	Oceania-Polynesia
Location (longitude, latitude)	13°48'S 171°47'W[a]
Currency	tala
Population (1995 est., in thousands)	156
Surface area (square kms)	2831
Population density (pop. per square km)	60
Sex ratio (females per 100 males)	93
Largest city (pop. in thousands)	Apia (35)[a]
UN membership date	15-Dec-1976
Major language(s)	Samoan, English

Economic indicators	1985	1994
Exchange rate (US$)	2.31	2.50
Consumer price index (1980=100)	203	382[b]
Tourist arrivals (in thousands)	49[c]	38[d]
GDP (million US$)	85	120[e]
GDP (per capita US$)	543	759[e]
Long-term rate of change in GDP (% per annum)	6.0	10.8[e]
Economically active female population (%)	53.8[f]	49.1
Economically active male population (%)	83.1[f]	83.6
Agricultural production index (1979-1981=100)	106[c]	87[d]
Food production index (1979-1981=100)	106[c]	87[d]
Commercial energy production (1,000 Mt coal equiv)	1[f]	2[e]
Motor vehicles (per 1,000 population)	29	...
Telephone lines (per 100 inhabitants)	2.4[g]	2.5[e]

Social indicators	1990/95
Age group 0-14 years (%)	44.3[f]
Age group 60+ years (women and men, %)	2.5/2.4[h]
Life expectancy at birth (women and men, years)	.../61
Infant mortality rate (per 1,000 births)	33[i]
Contraceptive use (% of currently married women)	34[i]
Urban population (%)	23
Urban population growth rate (% per annum)	1.4
Rural population growth rate (% per annum)	-0.2
Foreign-born (1985, %)	3.0
Television receivers (per 1,000 population)	40
Parliamentary seats (women and men, %)	4/96[g]

Environment	1990/91
Threatened species	10
Forested area (%)	47.2
CO_2 emissions (10,000 Mt)	34
Energy consumption (1,000 Mt coal equiv)	399
Precipitation (mm)	2928[g]

a Capital city. b February 1994. c 1988. d 1992. e 1991. f 1980. g 1987. h 1981. i 1985. j Source: NATHR.

San Marino

Region	Southern Europe
Location (longitude, latitude)	43°94'N 12°43'E[a]
Currency	Italian lire
Surface area (square kms)	61
Population density (pop. per square km)	377
Largest city (pop. in thousands)	San Marino (4)[a]
UN membership date	02-Mar-1992
Major language(s)	Italian

Economic indicators	1985	1994
Unemployment (%)	...	4.5
Tourist arrivals (in thousands)	504[b]	583[c]
GDP (million US$)	163	457[d]
GDP (per capita US$)	7429	19857[d]
Long-term rate of change in GDP (% per annum)	2.4	1.7[d]
Labour force in industry (%)	49.0[e]	41.9[c]
Labour force in agriculture (%)	5.2[e]	2.2[c]
Telephone lines (per 100 inhabitants)	37.4[f]	47.0[d]

Social indicators	1990/95
Urban population (%)	94
Urban population growth rate (% per annum)	0.6
Rural population growth rate (% per annum)	−7.6
Foreign-born (1985,%)	38.6
Newspaper circulation (per 1,000 population)	87
Television receivers (per 1,000 population)	351

a Capital city. b 1988. c 1992. d 1991. e 1980. f 1987.

Sao Tome and Principe

Region	Middle Africa
Location (longitude, latitude)	0°20'N 6°44'E[a]
Currency	dobra
Population (1995 est., in thousands)	97
Surface area (square kms)	964
Population density (pop. per square km)	129
Sex ratio (females per 100 males)	101
Largest city (pop. in thousands)	Sao Tome (50)[a]
UN membership date	16-Sep-1975
Major language(s)	Portuguese

Economic indicators	1985	1994
Exchange rate (US$)	41.20	813.66
Balance of payments, current account (million US$)	−18	−14[b]
Tourist arrivals (in thousands)	1[c]	1[d]
GDP (million US$)	37	45[e]
GDP (per capita US$)	349	373[e]
Long-term rate of change in GDP (% per annum)	−1.6	3.5[e]
Agricultural production index (1979-1981=100)	96[c]	79[d]
Food production index (1979-1981=100)	96[c]	79[d]
Commercial energy production (1,000 Mt coal equiv)	1[f]	1[e]
Motor vehicles (per 1,000 population)	23	...
Telephone lines (per 100 inhabitants)	1.9[g]	1.8[e]

Social indicators	1990/95
Age group 0-14 years (%)	46.3[f]
Age group 60+ years (women and men, %)	3.9/3.2[h]
Infant mortality rate (per 1,000 births)	72
Total fertility rate (births per woman)	6
Urban population (%)	47
Urban population growth rate (% per annum)	4.2
Rural population growth rate (% per annum)	0.6
Foreign-born (1985,%)	6.7
Parliamentary seats (women and men, %)	11/89

Environment	1990/91
Threatened species	15
CO_2 emissions (10,000 Mt)	19
Energy consumption (1,000 Mt coal equiv)	289
Precipitation (mm)	872[g]
Average temperature (January and July, centigrade)	27.6/23.8[f]

a Capital city. b 1990. c 1988. d 1992. e 1991. f 1980. g 1987. h 1981.

Saudi Arabia

Region	Western Asia
Location (longitude, latitude)	24°63'N 46°69'E
Currency	Saudi Arabian riyal
Population (1995 est., in thousands)	17608
Surface area (square kms)	2149690
Population density (pop. per square km)	7
Sex ratio (females per 100 males)	81
Largest city (pop. in thousands)	Riyadh (1975)
UN membership date	24-Oct-1945
Major language(s)	Arabic

Economic indicators	1985	1994
Exchange rate (US$)	3.64	3.75
Consumer price index (1980=100)	97	101[a]
Balance of payments, current account (million US$)	–12932	–19431[b]
Tourist arrivals (in thousands)	763[c]	750[b]
GDP (million US$)	97017	90059[d]
GDP (per capita US$)	7837	5853[d]
Long-term rate of change in GDP (% per annum)	–8.0	6.0
Gross fixed capital formation (% of GDP)	24.3	...
Economically active female population (%)	7.0[e]	9.2
Economically active male population (%)	84.8[e]	84.3
Annual growth of econ. active female pop. (%)	6.9[f]	...
Annual growth of econ. active male pop. (%)	4.8[f]	...
Agricultural production index (1979-1981=100)	511[c]	583[b]
Food production index (1979-1981=100)	521[c]	595[b]
Commercial energy production (1,000 Mt coal equiv)[g]	728427[e]	660709[a]
Motor vehicles (per 1,000 population)	222	202[f]
Telephone lines (per 100 inhabitants)	7.5[h]	8.4[d]

Social indicators	1990/95
Growth rate of population (% per annum)	3.4
Age group 0-14 years (%)	42.0
Age group 60+ years (women and men, %)	2.1/2.2
Life expectancy at birth (women and men, years)	71/68
Infant mortality rate (per 1,000 births)	31
Total fertility rate (births per woman)	6
Urban population (%)	80
Urban population growth rate (% per annum)	4.1
Rural population growth rate (% per annum)	0.6
Foreign-born (1985, %)	29.2
Refugees	28700
Government education expenditure (% of GDP)	6[i]
Primary-secondary gross enrolment ratio (f and m, per 100)	59/71[i]
Third-level students (per 100,000 population)	1089
Newspaper circulation (per 1,000 population)	40
Television receivers (per 1,000 population)	266

Environment	1990/91
Threatened species	28
Forested area (%)	0.6
CO_2 emissions (10,000 Mt)	586657[g]
Energy consumption (1,000 Mt coal equiv)[g]	6922
Precipitation (mm)	82[h]
Average temperature (January and July, centigrade)	16.1/40.1[e]

a February 1994. b 1992. c 1988. d 1991. e 1980. f 1990. g Includes part of the Neutral Zone. h 1987. i 1989.

Senegal

Region	Western Africa
Location (longitude, latitude)	14°83'N 17°67'W
Currency	CFA franc
Population (1995 est., in thousands)	8387
Surface area (square kms)	196722
Population density (pop. per square km)	38
Sex ratio (females per 100 males)	100
Largest city (pop. in thousands)	Dakar (1613)
UN membership date	28-Sep-1960
Major language(s)	French, Wolof, Ful, Serer, Dyola, Malinke

Economic indicators	1985	1994
Exchange rate (US$)	378.05	528.15
Consumer price index (1980=100)	175	173[a]
Industrial production index (1980 = 100)	118	106[bc]
Balance of payments, current account (million US$)	−273	−238[d]
Tourist arrivals (in thousands)	256[e]	246[f]
GDP (million US$)	2578	5608[d]
GDP (per capita US$)	404	745[d]
Long-term rate of change in GDP (% per annum)	3.8	2.6[d]
Economically active female population (%)	58.7[g]	51.5
Economically active male population (%)	87.2[g]	85.6
Annual growth of econ. active female pop. (%)	2.4[h]	...
Annual growth of econ. active male pop. (%)	2.9[h]	...
Agricultural production index (1979-1981=100)	127[e]	139[f]
Food production index (1979-1981=100)	126[e]	139[f]
Motor vehicles (per 1,000 population)	16	...
Telephone lines (per 100 inhabitants)	0.4[i]	0.6[d]

Social indicators	1990/95
Growth rate of population (% per annum)	2.7
Age group 0-14 years (%)	44.6
Age group 60+ years (women and men, %)	2.5/2.2
Life expectancy at birth (women and men, years)	50/48
Infant mortality rate (per 1,000 births)	80
Total fertility rate (births per woman)	6
Contraceptive use (% of currently married women)	7
Urban population (%)	42
Urban population growth rate (% per annum)	3.9
Rural population growth rate (% per annum)	1.9
Foreign-born (1985,%)	1.0
Refugees	71600
Government education expenditure (% of GDP)	4[e]
Primary-secondary gross enrolment ratio (f and m, per 100)	31/45[j]
Third-level students (per 100,000 population)	253[j]
Newspaper circulation (per 1,000 population)	7
Television receivers (per 1,000 population)	36
Intentional homicides (1986, per 100,000 population)	2
Parliamentary seats (women and men, %)	12/88

Environment	1990/91
Threatened species	27
Forested area (%)	53.6
CO_2 emissions (10,000 Mt)	764
Energy consumption (1,000 Mt coal equiv)	147
Precipitation (mm)	578[i]
Average temperature (January and July, centigrade)	21.1/27.3[g]

a December 1993. b Provisional or estimated figure. c June 1993.
d 1991. e 1988. f 1992. g 1980. h 1990. i 1987. j 1989.

Seychelles

Region	Eastern Africa
Location (longitude, latitude)	4°35'S 55°28'E[a]
Currency	Seychelles rupee
Population (1995 est., in thousands)	67
Surface area (square kms)	455
Population density (pop. per square km)	149
Sex ratio (females per 100 males)	100
Largest city (pop. in thousands)	Victoria (42)[a]
UN membership date	21-Sep-1976
Major language(s)	Seychellois

Economic indicators	1985	1994
Exchange rate (US$)	6.60	4.93
Consumer price index (1980=100)	122	108[b]
Balance of payments, current account (million US$)	−19	−2[c]
Tourist arrivals (in thousands)	77[d]	99[c]
GDP (million US$)	169	387[e]
GDP (per capita US$)	2521	5447[e]
Long-term rate of change in GDP (% per annum)	10.3	0.0[e]
Gross fixed capital formation (% of GDP)	22.7	22.7[f]
Labour force in industry (%)	25.1[g]	18.8[h]
Labour force in agriculture (%)	10.6[g]	9.9[h]
Motor vehicles (per 1,000 population)	73	82[f]
Telephone lines (per 100 inhabitants)	9.9[i]	11.8[e]

Social indicators	1990/95
Growth rate of population (% per annum)	0.4
Age group 0-14 years (%)	35.1
Age group 60+ years (women and men, %)	5.5/3.4[hi]
Life expectancy at birth (women and men, years)	74/65[g]
Infant mortality rate (per 1,000 births)	18
Urban population (%)	65
Urban population growth rate (% per annum)	2.8
Rural population growth rate (% per annum)	−2.2
Foreign-born (1985,%)	2.4
Government education expenditure (% of GDP)	9
Newspaper circulation (per 1,000 population)	45
Television receivers (per 1,000 population)	85
Intentional homicides (1986, per 100,000 population)	5
Parliamentary seats (women and men, %)	46/54

Environment	1990/91
Threatened species	22
Forested area (%)	17.9
CO_2 emissions (10,000 Mt)	36
Energy consumption (1,000 Mt coal equiv)	887

a Capital city. b January 1994. c 1992. d 1988. e 1991. f 1990. g 1980. h 1989. i 1987. j Source DYB90.

Sierra Leone

Region	Western Africa
Location (longitude, latitude)	8°30'N 13°15'W[a]
Currency	leone
Population (1995 est., in thousands)	4740
Surface area (square kms)	71740
Population density (pop. per square km)	59
Sex ratio (females per 100 males)	103
Largest city (pop. in thousands)	Freetown (669)[a]
UN membership date	27-Sep-1961
Major language(s)	English, Mende, Temne, Krio, Limba

Economic indicators	1985	1994
Exchange rate (US$)	5.21	597.12
Consumer price index (1980=100)	776	58824
Balance of payments, current account (million US$)	4	−69[b]
Tourist arrivals (in thousands)	75[c]	91[d]
GDP (million US$)	939	508[e]
GDP (per capita US$)	256	119[e]
Long-term rate of change in GDP (% per annum)	7.5	−2.6[e]
Gross fixed capital formation (% of GDP)	9.8	10.1[b]
Economically active female population (%)	41.4[f]	36.9
Economically active male population (%)	84.6[f]	82.5
Annual growth of econ. active female pop. (%)	0.8[b]	...
Annual growth of econ. active male pop. (%)	1.4[b]	...
Labour force in industry (%)	34.2[f]	34.8[c]
Labour force in agriculture (%)	9.2[f]	10.4[c]
Agricultural production index (1979-1981=100)	122[c]	118[d]
Food production index (1979-1981=100)	118[c]	113[d]
Motor vehicles (per 1,000 population)	12	...
Telephone lines (per 100 inhabitants)	0.4[g]	0.4[e]

Social indicators	1990/95
Growth rate of population (% per annum)	2.7
Age group 0-14 years (%)	44.9
Age group 60+ years (women and men, %)	2.8/2.3
Life expectancy at birth (women and men, years)	45/41
Infant mortality rate (per 1,000 births)	143
Total fertility rate (births per woman)	7
Contraceptive use (% of currently married women)	4[h]
Urban population (%)	36
Urban population growth rate (% per annum)	5.0
Rural population growth rate (% per annum)	1.4
Foreign-born (1985,%)	4.3
Refugees	5900
Government education expenditure (% of GDP)	1[i]
Primary-secondary gross enrolment ratio (f and m, per 100)	27/41
Third-level students (per 100,000 population)	125[i]
Newspaper circulation (per 1,000 population)	2
Television receivers (per 1,000 population)	10

Environment	1990/91
Threatened species	29
Forested area (%)	28.7
CO_2 emissions (10,000 Mt)	188
Energy consumption (1,000 Mt coal equiv)	76
Precipitation (mm)	3318[j]
Average temperature (January and July, centigrade)	26.6/25.1[f]

a Capital city. b 1990. c 1988. d 1992. e 1991. f 1980. g 1987. h Source: UNICEF94. i 1989. j Lungi.

Singapore

Region	South-eastern Asia
Location (longitude, latitude)	1°29'N 103°85'E
Currency	Singapore dollar
Population (1995 est., in thousands)	2853
Surface area (square kms)	618
Population density (pop. per square km)	4471
Sex ratio (females per 100 males)	97
Largest city (pop. in thousands)	Singapore (2710)
UN membership date	21-Sep-1965
Major language(s)	Malay, English, Chinese, Tamil

Economic indicators	1985	1994
Exchange rate (US$)	2.11	1.48
Consumer price index (1980=100)	117	141[a]
Balance of payments, current account (million US$)	–4	2929[b]
Tourist arrivals (in thousands)	3833[c]	5446[b]
GDP (million US$)	17693	39984[d]
GDP (per capita US$)	6917	14598[d]
Long-term rate of change in GDP (% per annum)	–1.6	6.7[d]
Gross fixed capital formation (% of GDP)	42.2	39.9[d]
Economically active female population (%)	44.5[e]	46.0
Economically active male population (%)	81.5[e]	79.7
Annual growth of econ. active female pop. (%)	4.1[f]	...
Annual growth of econ. active male pop. (%)	2.5[f]	...
Labour force in industry (%)	35.8[e]	34.6[b]
Labour force in agriculture (%)	1.3[e]	0.3[b]
Agricultural production index (1979-1981=100)	76[c]	56[b]
Food production index (1979-1981=100)	76[c]	56[b]
Motor vehicles (per 1,000 population)	139	159[d]
Telephone lines (per 100 inhabitants)	33.8[g]	40.2[d]

Largest export industries		Major trading partners		1992	
	(% of exports)	(% of exports)		(% of imports)	
Metal manufacture	60	United States	21	Japan	21
Chemicals	21	Malaysia	13	United States	17
Textiles	5	Hong Kong	8	Malaysia	15

Social indicators	1990/95
Growth rate of population (% per annum)	1.0
Age group 0-14 years (%)	22.6
Age group 60+ years (women and men, %)	5.2/4.5
Life expectancy at birth (women and men, years)	77/72
Infant mortality rate (per 1,000 births)	8
Total fertility rate (births per woman)	2
Urban population (%)	100
Urban population growth rate (% per annum)	1.0
Foreign-born (1985,%)	18.6
Refugees	100
Government education expenditure (% of GDP)	3[c]
Primary-secondary gross enrolment ratio (f and m, per 100)	87/87[h]
Third-level students (per 100,000 population)	963[e]
Newspaper circulation (per 1,000 population)	282
Television receivers (per 1,000 population)	378
Intentional homicides (1986, per 100,000 population)	3
Parliamentary seats (women and men, %)	4/96

Environment	1990/91
Threatened species	21
Forested area (%)	4.8
CO_2 emissions (10,000 Mt)	11270
Energy consumption (1,000 Mt coal equiv)	5821
Precipitation (mm)	2282[g]
Average temperature (January and July, centigrade)	26.1/27.4[e]

a August 1994. b 1992. c 1988. d 1991. e 1980. f 1990. g 1987. h 1989.

Slovakia

Region Eastern Europe
Location (longitude, latitude) 48°09'N 17°07'E
Currency koruna
Largest city (pop. in thousands) Bratislava (441)[a]
UN membership date 19-Jan-1993
Major language(s) Slovak

Economic indicators	1985	1994
Exchange rate (US$)	...	31.00
GDP (million US$)	11859	9878[b]
GDP (per capita US$)	2297	1872[b]
Long-term rate of change in GDP (% per annum)	4.1	0.0[b]
Gross fixed capital formation (% of GDP)	29.4	28.0[b]
Economically active female population (%)	59.5[c]	62.1
Economically active male population (%)	76.0[c]	76.4

Social indicators	1990/95
Contraceptive use (% of currently married women)	74

Environment	1990/91
Threatened species	44

a Capital city. b 1991. c 1980.

Slovenia

Region Southern Europe
Location (longitude, latitude) 46°05'N 14°51'E
Currency koruna
Largest city (pop. in thousands) Ljubljana (277)[a]
UN membership date 22-May-1992
Major language(s) Slovenian

Economic indicators	1985	1994
Consumer price index (1980=100)[b]	72	484[c]
Balance of payments, current account (million US$)	1181[d]	...
GDP (million US$)	7678	17856[e]
GDP (per capita US$)	3989	9096[e]
Long-term rate of change in GDP (% per annum)	1.1	−9.3[e]
Economically active female population (%)	70.7[f]	70.3
Economically active male population (%)	82.4[f]	77.0
Labour force in industry (%)	52.5[g]	48.4[d]
Labour force in agriculture (%)	9.5[g]	9.2[d]
Telephone lines (per 100 inhabitants)	21.9[h]	23.2[e]

Social indicators	1990/95
Refugees	47000
Television receivers (per 1,000 population)	284
Intentional homicides (1986, per 100,000 population)	3[i]

Environment	1990/91
Threatened species	5

a Capital city. b Multiply each figure by 10. c May 1994. d 1992.
e 1991. f 1980. g 1983. h 1990. i Source: DYB92.

Solomon Islands

Region	Oceania-Melanesia
Location (longitude, latitude)	9°43'S 159°93'E[a]
Currency	Solomon Islands dollar
Population (1995 est., in thousands)	378
Surface area (square kms)	28896
Population density (pop. per square km)	11
Sex ratio (females per 100 males)	94
Largest city (pop. in thousands)	Honiara (39)[a]
UN membership date	19-Sep-1978
Major language(s)	English, Melanesian languages, Papuan languages

Economic indicators	1985	1994
Exchange rate (US$)	1.61	3.29
Consumer price index (1980=100)	171	509[b]
Balance of payments, current account (million US$)	−21	−37[c]
Tourist arrivals (in thousands)	11[d]	12[e]
GDP (million US$)	160	195[c]
GDP (per capita US$)	593	589[c]
Long-term rate of change in GDP (% per annum)	2.8	4.0[c]
Gross fixed capital formation (% of GDP)	21.1	...
Economically active female population (%)	55.7[f]	51.0
Economically active male population (%)	86.5[f]	85.6
Agricultural production index (1979-1981=100)	114[d]	142[e]
Food production index (1979-1981=100)	114[d]	143[e]
Telephone lines (per 100 inhabitants)	1.0[g]	1.4[c]

Social indicators	1990/95
Growth rate of population (% per annum)	3.3
Age group 0-14 years (%)	44.4
Age group 60+ years (women and men, %)	2.1/2.1
Life expectancy at birth (women and men, years)	73/68
Infant mortality rate (per 1,000 births)	27
Total fertility rate (births per woman)	5
Contraceptive use (% of currently married women)	3[hi]
Urban population (%)	17
Urban population growth rate (% per annum)	6.5
Rural population growth rate (% per annum)	2.7
Foreign-born (1985,%)	2.5
Government education expenditure (% of GDP)	5
Primary-secondary gross enrolment ratio (f and m, per 100)	48/61[j]

Environment	1990/91
Threatened species	39
Forested area (%)	88.6
CO_2 emissions (10,000 Mt)	44
Energy consumption (1,000 Mt coal equiv)	233
Precipitation (mm)	2097[g]
Average temperature (January and July, centigrade)	26.8/26.2[f]

a Capital city. b July 1994. c 1991. d 1988. e 1992. f 1980. g 1987. h 1989. i Source: NATHR. j 1986.

Somalia

Region	Eastern Africa
Location (longitude, latitude)	2°01'N 45°37'E
Currency	Somali shilling
Population (1995 est., in thousands)	10173
Surface area (square kms)	637657
Population density (pop. per square km)	12
Sex ratio (females per 100 males)	102
Largest city (pop. in thousands)	Mogadishu (779)
UN membership date	20-Sep-1960
Major language(s)	Somali, Arabic

Economic indicators	1985	1994
Exchange rate (US$)	42.50	4900.00
Consumer price index (1980=100)	640	...
Balance of payments, current account (million US$)	−103	...
Tourist arrivals (in thousands)	40[a]	20[b]
GDP (million US$)	2211	1253[c]
GDP (per capita US$)	281	141[c]
Long-term rate of change in GDP (% per annum)	9.5	−1.0[c]
Gross fixed capital formation (% of GDP)	8.9	...
Economically active female population (%)	57.6[d]	51.2
Economically active male population (%)	88.7[d]	86.7
Annual growth of econ. active female pop. (%)	2.5[e]	...
Annual growth of econ. active male pop. (%)	3.0[e]	...
Agricultural production index (1979-1981=100)	124[a]	45[b]
Food production index (1979-1981=100)	124[a]	45[b]
Motor vehicles (per 1,000 population)	2	...
Telephone lines (per 100 inhabitants)	0.2[f]	0.2[c]

Social indicators	1990/95
Growth rate of population (% per annum)	3.2
Age group 0-14 years (%)	47.5
Age group 60+ years (women and men, %)	2.3/2.0
Life expectancy at birth (women and men, years)	49/45
Infant mortality rate (per 1,000 births)	122
Total fertility rate (births per woman)	7
Contraceptive use (% of currently married women)	1
Urban population (%)	26
Urban population growth rate (% per annum)	4.4
Rural population growth rate (% per annum)	2.8
Foreign-born (1985,%)	9.3
Refugees	500
Government education expenditure (% of GDP)	0[g]
Primary-secondary gross enrolment ratio (f and m, per 100)	9/18[h]
Third-level students (per 100,000 population)	195[g]
Newspaper circulation (per 1,000 population)	1
Television receivers (per 1,000 population)	12
Parliamentary seats (women and men, %)	4/96[f]

Environment	1990/91
Threatened species	31
Forested area (%)	14.2
CO_2 emissions (10,000 Mt)	143
Energy consumption (1,000 Mt coal equiv)	10
Precipitation (mm)	422[f]
Average temperature (January and July, centigrade)	26.5/26.0[d]

a 1988. b 1992. c 1991. d 1980. e 1990. f 1987. g 1986. h 1985.

South Africa

Region	Western Africa
Location (longitude, latitude)	33°92'S 18°44'E
Currency	rand
Population (1995 est., in thousands)	42741
Surface area (square kms)	1221037
Population density (pop. per square km)	30
Sex ratio (females per 100 males)	101
Largest city (pop. in thousands)	Cape Town (2297)
UN membership date	07-Nov-1945
Major language(s)	Afrikaans, English, Zulu, Xhosa, Tswana, Pedi

Economic indicators	1985	1994
Exchange rate (US$)	2.56	3.56
Consumer price index (1980=100)	193	627[a]
Balance of payments, current account (million US$)	2622	1814[b]
Tourist arrivals (in thousands)	805[c]	2892[d]
GDP (million US$)	56196	108076[e]
GDP (per capita US$)	1673	2780[e]
Long-term rate of change in GDP (% per annum)	−1.2	−0.4[e]
Gross fixed capital formation (% of GDP)	23.3	18.0[e]
Economically active female population (%)	37.7[f]	40.6
Economically active male population (%)	76.3[f]	75.6
Annual growth of econ. active female pop. (%)	2.5[g]	...
Annual growth of econ. active male pop. (%)	1.9[g]	...
Labour force in industry (%)	52.8[f]	48.1[d]
Agricultural production index (1979-1981=100)	106[c]	84[d]
Food production index (1979-1981=100)	106[c]	83[d]
Motor vehicles (per 1,000 population)	136	137[e]
Telephone lines (per 100 inhabitants)	7.6[h]	8.8[e]

Largest export industries	Major trading partners			1992
(% of exports)	(% of exports)[i]		(% of imports)[i]	
...	Switzerland[j]	12	Germany	16
...	United Kingdom	8	United States	14
...	Japan	7	Japan	11

Social indicators	1990/95
Growth rate of population (% per annum)	2.4
Age group 0-14 years (%)	37.5
Age group 60+ years (women and men, %)	3.5/2.7
Life expectancy at birth (women and men, years)	66/60
Infant mortality rate (per 1,000 births)	53
Total fertility rate (births per woman)	4
Contraceptive use (% of currently married women)	50[c]
Urban population (%)	51
Urban population growth rate (% per annum)	3.0
Rural population growth rate (% per annum)	1.7
Foreign-born (1985,%)	5.5
Newspaper circulation (per 1,000 population)	35
Television receivers (per 1,000 population)	98
Intentional homicides (1986, per 100,000 population)	26
Parliamentary seats (women and men, %)	3/97

Environment	1990/91
Threatened species	230
Forested area (%)	3.7
CO_2 emissions (10,000 Mt)	76063
Precipitation (mm)	746[k]
Average temperature (January and July, centigrade)	21.0/10.3[l]

a August 1994. b 1993. c 1988. d 1992. e 1991. f 1980. g 1990. h 1987. i South African Customs Union. j Includes Liechtenstein. k Pretoria.

Spain

Region	Southern Europe
Location (longitude, latitude)	40°44'N 3°68'W
Currency	peseta
Population (1995 est., in thousands)	39276
Surface area (square kms)	504782
Population density (pop. per square km)	77
Sex ratio (females per 100 males)	103
Largest city (pop. in thousands)	Madrid (5234)
UN membership date	14-Dec-1955
Major language(s)	Spanish, Catalan

Economic indicators	1985	1994
Exchange rate (US$)	154.15	128.29
Consumer price index (1980=100)	178	301[a]
Industrial production index (1980 = 100)	103	119[b]
Balance of payments, current account (million US$)	2851	−6257[c]
Tourist arrivals (in thousands)	38784[d]	39638[e]
GDP (million US$)	165849	527294[b]
GDP (per capita US$)	4311	13511[b]
Long-term rate of change in GDP (% per annum)	2.6	2.3[b]
Gross fixed capital formation (% of GDP)	19.2	24.0[b]
Economically active female population (%)	21.4[f]	22.0
Economically active male population (%)	74.3[f]	71.4
Annual growth of econ. active female pop. (%)	2.2[g]	...
Annual growth of econ. active male pop. (%)	0.7[g]	...
Labour force in industry (%)	35.9[f]	32.4[e]
Labour force in agriculture (%)	19.3[f]	10.1[e]
Agricultural production index (1979-1981=100)	120[d]	123[e]
Food production index (1979-1981=100)	119[d]	123[e]
Commercial energy production (1,000 Mt coal equiv)	22331[f]	44800[b]
Motor vehicles (per 1,000 population)	283	389[b]
Telephone lines (per 100 inhabitants)	26.4[h]	34.0[b]

Largest export industries		Major trading partners			1992
	(% of exports)		(% of exports)		(% of imports)
Metal manufacture	49	France[i]	20	Germany	16
Chemicals	14	Germany	16	France[i]	16
Agriculture	9	Italy	11	Italy	10

Social indicators	1990/95
Growth rate of population (% per annum)	0.2
Age group 0-14 years (%)	17.0
Age group 60+ years (women and men, %)	11.4/8.8
Life expectancy at birth (women and men, years)	80/75
Infant mortality rate (per 1,000 births)	7
Total fertility rate (births per woman)	1
Contraceptive use (% of currently married women)	59[j]
Urban population (%)	81
Urban population growth rate (% per annum)	0.7
Rural population growth rate (% per annum)	−2.0
Foreign-born (1985,%)	1.0
Refugees	9700
Government education expenditure (% of GDP)	4[k]
Primary-secondary gross enrolment ratio (f and m, per 100)	111/106[k]
Third-level students (per 100,000 population)	2981[k]
Newspaper circulation (per 1,000 population)	82[d]
Television receivers (per 1,000 population)	400
Intentional homicides (1986, per 100,000 population)	2
Parliamentary seats (women and men, %)	16/84

Environment	1990/91
Threatened species	136
Forested area (%)	31.0
CO_2 emissions (10,000 Mt)	60010
Energy consumption (1,000 Mt coal equiv)	3065
Precipitation (mm)	436[h]
Average temperature (January and July, centigrade)	4.9/24.2[f]

a August 1994. b 1991. c 1993. d 1988. e 1992. f 1980. g 1990. h 1987.
i Includes Monaco. j Source: UNICEF94. k 1989.

Sri Lanka

Region	Southern Asia
Location (longitude, latitude)	6°92'N 79°85'E
Currency	Sri Lanka rupee
Population (1995 est., in thousands)	18346
Surface area (square kms)	65610
Population density (pop. per square km)	263
Sex ratio (females per 100 males)	101
Largest city (pop. in thousands)	Colombo (616)[a]
UN membership date	14-Dec-1955
Major language(s)	Sinhalese, Tamil

Economic indicators	1985	1994
Exchange rate (US$)	27.41	49.53
Consumer price index (1980=100)	176	461
Balance of payments, current account (million US$)	−419	−451[b]
Tourist arrivals (in thousands)	183[c]	394[b]
GDP (million US$)	5808	8936[d]
GDP (per capita US$)	361	512[d]
Long-term rate of change in GDP (% per annum)	5.0	4.8[d]
Gross fixed capital formation (% of GDP)	23.9	23.4[d]
Economically active female population (%)	30.7[e]	28.8
Economically active male population (%)	80.8[e]	79.3
Annual growth of econ. active female pop. (%)	2.4[f]	...
Annual growth of econ. active male pop. (%)	1.9[f]	...
Labour force in industry (%)	26.8[e]	36.2[f]
Labour force in agriculture (%)	49.8[e]	42.0[f]
Agricultural production index (1979-1981=100)	99[c]	94[b]
Food production index (1979-1981=100)	101[c]	98[b]
Commercial energy production (1,000 Mt coal equiv)	182[e]	383[d]
Motor vehicles (per 1,000 population)	17	19[d]
Telephone lines (per 100 inhabitants)	0.6[g]	0.7[d]

Largest export industries		Major trading partners			1992
	(% of exports)		(% of exports)		(% of imports)
Textiles	53	United States	34	Japan	12
Agriculture	26	Germany	9	India	9
Chemicals	4	United Kingdom	7	Hong Kong	7

Social indicators	1990/95
Growth rate of population (% per annum)	1.3
Age group 0-14 years (%)	30.3
Age group 60+ years (women and men, %)	4.4/4.3
Life expectancy at birth (women and men, years)	74/70
Infant mortality rate (per 1,000 births)	24
Total fertility rate (births per woman)	2
Contraceptive use (% of currently married women)	62[g]
Urban population (%)	22
Urban population growth rate (% per annum)	2.2
Rural population growth rate (% per annum)	1.0
Foreign-born (1985, %)	0.2
Government education expenditure (% of GDP)	2
Primary-secondary gross enrolment ratio (f and m, per 100)	89/87
Third-level students (per 100,000 population)	400[c]
Newspaper circulation (per 1,000 population)	32
Television receivers (per 1,000 population)	981
Intentional homicides (1986, per 100,000 population)	13
Parliamentary seats (women and men, %)	5/95

Environment	1990/91
Threatened species	51
Forested area (%)	31.7
CO_2 emissions (10,000 Mt)	1137
Energy consumption (1,000 Mt coal equiv)	133
Precipitation (mm)	2397[g]
Average temperature (January and July, centigrade)	26.2/27.1[e]

a Capital city. b 1992. c 1988. d 1991. e 1980. f 1990. g 1987.

Sudan

Region	Northern Africa
Location (longitude, latitude)	15°59'N 32°51'E
Currency	Sudanese pound
Population (1995 est., in thousands)	28960
Surface area (square kms)	2505813
Population density (pop. per square km)	10
Sex ratio (females per 100 males)	99
Largest city (pop. in thousands)	Khartoum (1953)
UN membership date	12-Nov-1956
Major language(s)	Arabic, Dinka

Economic indicators	1985	1994
Exchange rate (US$)	2.50	387.60
Consumer price index (1980=100)	398	385[a]
Balance of payments, current account (million US$)	148	−506[b]
Tourist arrivals (in thousands)	37[c]	17[b]
GDP (million US$)	6081	15927[d]
GDP (per capita US$)	279	614[d]
Long-term rate of change in GDP (% per annum)	−2.9	0.0[d]
Gross fixed capital formation (% of GDP)	24.6	...
Economically active female population (%)	21.1[e]	25.9
Economically active male population (%)	87.8[e]	85.8
Annual growth of econ. active female pop. (%)	3.3[a]	...
Annual growth of econ. active male pop. (%)	2.8[a]	...
Labour force in industry (%)	32.9[f]	19.6[a]
Labour force in agriculture (%)	12.8[f]	4.8[b]
Agricultural production index (1979-1981=100)	117[c]	131[b]
Food production index (1979-1981=100)	118[c]	134[b]
Commercial energy production (1,000 Mt coal equiv)	83[e]	115[d]
Motor vehicles (per 1,000 population)	9	...
Telephone lines (per 100 inhabitants)	0.3[g]	0.2[d]

Social indicators	1990/95
Growth rate of population (% per annum)	2.8
Age group 0-14 years (%)	44.5
Age group 60+ years (women and men, %)	2.5/2.2
Life expectancy at birth (women and men, years)	53/51
Infant mortality rate (per 1,000 births)	99
Total fertility rate (births per woman)	6
Contraceptive use (% of currently married women)	9
Urban population (%)	25
Urban population growth rate (% per annum)	4.5
Rural population growth rate (% per annum)	2.2
Foreign-born (1985,%)	4.7
Refugees	725600
Government education expenditure (% of GDP)	4[h]
Primary-secondary gross enrolment ratio (f and m, per 100)	30/42[h]
Third-level students (per 100,000 population)	246[i]
Newspaper circulation (per 1,000 population)	24
Television receivers (per 1,000 population)	77
Parliamentary seats (women and men, %)	5/95

Environment	1990/91
Threatened species	36
Forested area (%)	17.9
CO_2 emissions (10,000 Mt)	929
Energy consumption (1,000 Mt coal equiv)	62
Precipitation (mm)	164[g]
Average temperature (January and July, centigrade)	22.5/30.8[e]

a 1990. b 1992. c 1988. d 1991. e 1980. f 1982. g 1987. h 1985. i 1989.

Suriname

Region	South America
Location (longitude, latitude)	5°83'N 55°17'W[a]
Currency	Suriname guilder
Population (1995 est., in thousands)	463
Surface area (square kms)	163265
Population density (pop. per square km)	3
Sex ratio (females per 100 males)	101
Largest city (pop. in thousands)	Paramaribo (114)[a]
UN membership date	04-Dec-1975
Major language(s)	Dutch, Sranan, Hindi, Javanese

Economic indicators	1985	1994
Exchange rate (US$)	1.78	1.79
Consumer price index (1980=100)	140	3223[b]
Balance of payments, current account (million US$)	−12	11[c]
Tourist arrivals (in thousands)	21[d]	30[c]
GDP (million US$)	978	1872[e]
GDP (per capita US$)	2555	4354[e]
Long-term rate of change in GDP (% per annum)	2.0	2.8[e]
Gross fixed capital formation (% of GDP)	17.9	18.7[e]
Economically active female population (%)	26.6[f]	30.8
Economically active male population (%)	71.4[f]	75.1
Annual growth of econ. active female pop. (%)	2.4[g]	...
Annual growth of econ. active male pop. (%)	1.2[g]	...
Labour force in industry (%)	21.8[h]	16.8[g]
Agricultural production index (1979-1981=100)	98[d]	102[c]
Food production index (1979-1981=100)	98[d]	102[c]
Commercial energy production (1,000 Mt coal equiv)	110[f]	471[e]
Motor vehicles (per 1,000 population)	116	128[e]
Telephone lines (per 100 inhabitants)	7.8[i]	9.4[e]

Social indicators	1990/95
Growth rate of population (% per annum)	1.9
Age group 0-14 years (%)	34.1
Age group 60+ years (women and men, %)	3.9/3.0
Life expectancy at birth (women and men, years)	73/68
Infant mortality rate (per 1,000 births)	28
Total fertility rate (births per woman)	3
Urban population (%)	50
Urban population growth rate (% per annum)	3.0
Rural population growth rate (% per annum)	0.8
Foreign-born (1985,%)	1.8
Refugees	100
Government education expenditure (% of GDP)	8
Primary-secondary gross enrolment ratio (f and m, per 100)	89/86[d]
Third-level students (per 100,000 population)	1025
Newspaper circulation (per 1,000 population)	95
Television receivers (per 1,000 population)	130
Intentional homicides (1986, per 100,000 population)	14
Parliamentary seats (women and men, %)	6/94

Environment	1990/91
Threatened species	17
Forested area (%)	91.0
CO_2 emissions (10,000 Mt)	551
Energy consumption (1,000 Mt coal equiv)	1879
Average temperature (January and July, centigrade)	26.4/27.1[f]

a Capital city. b February 1994. c 1992. d 1988. e 1991. f 1980. g 1990. h 1983. i 1987.

Swaziland

Region Western Africa
Location (longitude, latitude) 26°18'S 31°06'E[a]
Currency emalangeni
Population (1995 est., in thousands) 859
Surface area (square kms) 17364
Population density (pop. per square km) 47
Sex ratio (females per 100 males) 103
Largest city (pop. in thousands) Mbabane (47)[a]
UN membership date 24-Sep-1968
Major language(s) Swazi, English

Economic indicators	1985	1994
Exchange rate (US$)	2.56	3.56
Consumer price index (1980=100)	199	157[b]
Balance of payments, current account (million US$)	−39	25[c]
Tourist arrivals (in thousands)	196[c]	258[c]
GDP (million US$)	367	615[e]
GDP (per capita US$)	558	797[e]
Long-term rate of change in GDP (% per annum)	3.9	0.0[e]
Economically active female population (%)	57.6[f]	51.1
Economically active male population (%)	88.6[f]	85.7
Annual growth of econ. active female pop. (%)	1.8[g]	...
Annual growth of econ. active male pop. (%)	2.6[g]	...
Labour force in industry (%)	25.6[f]	26.2[h]
Labour force in agriculture (%)	39.9[f]	30.2[h]
Agricultural production index (1979-1981=100)	123[d]	115[c]
Food production index (1979-1981=100)	124[d]	116[c]
Motor vehicles (per 1,000 population)	60	71[g]
Telephone lines (per 100 inhabitants)	1.4[i]	1.9[e]

Social indicators	1990/95
Growth rate of population (% per annum)	2.7
Age group 0-14 years (%)	42.5
Age group 60+ years (women and men, %)	2.7/2.2
Life expectancy at birth (women and men, years)	60/56
Infant mortality rate (per 1,000 births)	73
Total fertility rate (births per woman)	5
Contraceptive use (% of currently married women)	20
Urban population (%)	31
Urban population growth rate (% per annum)	6.1
Rural population growth rate (% per annum)	1.3
Foreign-born (1985, %)	6.0
Refugees	55600
Government education expenditure (% of GDP)	6[j]
Primary-secondary gross enrolment ratio (f and m, per 100)	84/85[j]
Third-level students (per 100,000 population)	418
Newspaper circulation (per 1,000 population)	13
Television receivers (per 1,000 population)	19
Parliamentary seats (women and men, %)	2/98[i]

Environment	1990/91
Threatened species	11
Forested area (%)	6.0
CO_2 emissions (10,000 Mt)	90

a Capital city. b July 1994. c 1992. d 1988. e 1991. f 1980. g 1990.
h 1986. i 1987. j 1989.

Sweden

Region	Northern Europe
Location (longitude, latitude)	59°33'N 18°06'E
Currency	Swedish kronor
Population (1995 est., in thousands)	8773
Surface area (square kms)	449964
Population density (pop. per square km)	19
Sex ratio (females per 100 males)	102
Largest city (pop. in thousands)	Stockholm (1677)
UN membership date	19-Nov-1946
Major language(s)	Swedish

Economic indicators	1985	1994
Exchange rate (US$)	7.61	7.49
Consumer price index (1980=100)	154	251
Industrial production index (1980 = 100)	110	125[a]
Unemployment (%)	...	8.1
Balance of payments, current account (million US$)	−1231	−1836[b]
Tourist arrivals (in thousands)	830[c]	650[d]
GDP (million US$)	100626	236946[e]
GDP (per capita US$)	12051	27520[e]
Long-term rate of change in GDP (% per annum)	2.2	−1.8[e]
Gross fixed capital formation (% of GDP)	18.9	20.9[f]
Economically active female population (%)	53.6[g]	55.0
Economically active male population (%)	71.3[g]	71.4
Annual growth of econ. active female pop. (%)	1.9[f]	...
Annual growth of econ. active male pop. (%)	0.0[f]	...
Labour force in industry (%)	32.2[g]	26.6[d]
Labour force in agriculture (%)	5.6[g]	3.2[d]
Agricultural production index (1979-1981=100)	92[c]	86[d]
Food production index (1979-1981=100)	92[c]	86[d]
Commercial energy production (1,000 Mt coal equiv)	17124[g]	36501[e]
Motor vehicles (per 1,000 population)	444	499[e]
Telephone lines (per 100 inhabitants)	65.1[h]	68.7[e]

Largest export industries		Major trading partners			1992
	(% of exports)	(% of exports)		(% of imports)	
Metal manufacture	52	Germany	15	Germany	19
Paper, paper products	14	United Kingdom	10	United States	9
Basic metal industry	8	Norway[i]	8	United Kingdom	9

Social indicators	1990/95
Growth rate of population (% per annum)	0.5
Age group 0-14 years (%)	18.8
Age group 60+ years (women and men, %)	12.4/9.6
Life expectancy at birth (women and men, years)	81/75
Infant mortality rate (per 1,000 births)	6
Total fertility rate (births per woman)	2
Urban population (%)	85
Urban population growth rate (% per annum)	0.7
Rural population growth rate (% per annum)	−0.5
Foreign-born (1985,%)	8.5
Refugees	324500
Government education expenditure (% of GDP)	7
Primary-secondary gross enrolment ratio (f and m, per 100)	99/97
Third-level students (per 100,000 population)	2281
Newspaper circulation (per 1,000 population)	525
Television receivers (per 1,000 population)	468
Intentional homicides (1986, per 100,000 population)	2
Parliamentary seats (women and men, %)	34/66

Environment	1990/91
Threatened species	45
Forested area (%)	62.3
CO_2 emissions (10,000 Mt)	14601
Energy consumption (1,000 Mt coal equiv)	6892
Precipitation (mm)	555[h]
Average temperature (January and July, centigrade)	−2.9/17.8[g]

a August 1994. b 1993. c 1988. d 1992. e 1991. f 1990. g 1980. h 1987.
i Includes Svalbard and Jan Mayen Islands.

Switzerland

Region	Western Europe
Location (longitude, latitude)	47°37'N 8°54'E
Currency	Swiss franc
Population (1995 est., in thousands)	6955
Surface area (square kms)	41293
Population density (pop. per square km)	164
Sex ratio (females per 100 males)	104
Largest city (pop. in thousands)	Zurich (890)
Major language(s)	German, French, Italian, Rhaeto-Romance

Economic indicators	1985	1994
Exchange rate (US$)	2.08	1.29
Consumer price index (1980=100)	123	160
Industrial production index (1980 = 100)	103	132[a,b]
Unemployment (%)	...	4.4
Balance of payments, current account (million US$)	6040	13419[c]
Tourist arrivals (in thousands)	11700[d]	12800[c]
GDP (million US$)	92772	231998[e]
GDP (per capita US$)	14339	34304[e]
Long-term rate of change in GDP (% per annum)	3.7	−0.1[e]
Gross fixed capital formation (% of GDP)	23.8	25.5[e]
Economically active female population (%)	42.6[f]	42.8
Economically active male population (%)	79.4[f]	80.6
Annual growth of econ. active female pop. (%)	0.9[g]	...
Annual growth of econ. active male pop. (%)	0.1[g]	...
Labour force in industry (%)	38.1[f]	33.9
Labour force in agriculture (%)	6.9[f]	5.6
Agricultural production index (1979-1981=100)	107[d]	113[c]
Food production index (1979-1981=100)	107[d]	113[c]
Commercial energy production (1,000 Mt coal equiv)[h]	9061[f]	12635[c]
Motor vehicles (per 1,000 population)	437	503[e]
Telephone lines (per 100 inhabitants)	53.3[i]	60.3[e]

Largest export industries	(% of exports)[h]
Metal manufacture	49
Chemicals	25
Other manufacturing	6

Major trading partners			1992
(% of exports)[h]		(% of imports)[h]	
Germany	23	Germany	34
France[j]	10	France[j]	11
Italy	9	Italy	10

Social indicators	1990/95
Growth rate of population (% per annum)	0.7
Age group 0-14 years (%)	17.2
Age group 60+ years (women and men, %)	11.6/8.5
Life expectancy at birth (women and men, years)	81/75
Infant mortality rate (per 1,000 births)	7
Total fertility rate (births per woman)	2
Urban population (%)	64
Urban population growth rate (% per annum)	1.5
Rural population growth rate (% per annum)	−0.6
Foreign-born (1985, %)	17.2
Refugees	26700
Government education expenditure (% of GDP)	4
Primary-secondary gross enrolment ratio (f and m, per 100)	95/98
Third-level students (per 100,000 population)	2118
Newspaper circulation (per 1,000 population)	456
Television receivers (per 1,000 population)	406
Intentional homicides (1986, per 100,000 population)	2
Parliamentary seats (women and men, %)	18/83

Environment	1990/91
Threatened species	60
Forested area (%)	25.5
CO_2 emissions (10,000 Mt)	11420[h]
Energy consumption (1,000 Mt coal equiv)[h]	4698
Precipitation (mm)	1137[i]
Average temperature (January and July, centigrade)	−1.1/17.6[f]

a Provisional or estimated figure. b June 1994. c 1992. d 1988. e 1991.
f 1980. g 1990. h Includes Liechtenstein. i 1987. j Includes Monaco.

Syrian Arab Republic

Region	Western Asia
Location (longitude, latitude)	33°52'N 36°30'E
Currency	Syrian pound
Population (1995 est., in thousands)	14775
Surface area (square kms)	185180
Population density (pop. per square km)	70
Sex ratio (females per 100 males)	98
Largest city (pop. in thousands)	Damascus (1790)
UN membership date	24-Oct-1945
Major language(s)	Arabic

Economic indicators	1985	1994
Exchange rate (US$)	3.92	11.23
Consumer price index (1980=100)	184	1002a
Industrial production index (1980 = 100)	136	186b
Balance of payments, current account (million US$)	−958	...
Tourist arrivals (in thousands)	421d	55c
GDP (million US$)	21204	684c
GDP (per capita US$)	2049	27225e
GDP (per capita US$)	2049	2126e
Long-term rate of change in GDP (% per annum)	6.1	8.9e
Gross fixed capital formation (% of GDP)	21.1	16.1e
Economically active female population (%)	12.5f	16.4
Economically active male population (%)	77.8f	76.8
Annual growth of econ. active female pop. (%)	5.8b	...
Annual growth of econ. active male pop. (%)	3.4b	...
Labour force in industry (%)	29.2g	25.0e
Labour force in agriculture (%)	30.6h	28.2e
Agricultural production index (1979-1981=100)	137d	137c
Food production index (1979-1981=100)	137d	133c
Commercial energy production (1,000 Mt coal equiv)	13516f	40666e
Motor vehicles (per 1,000 population)	22	20e
Telephone lines (per 100 inhabitants)	4.2i	3.9e

Largest export industries	Major trading partners			1992
(% of exports)	(% of exports)		(% of imports)	
...	Italy	35	Germany	10
...	Francej	19	Japan	10
...	Lebanon	13	Italy	8

Social indicators	1990/95
Growth rate of population (% per annum)	3.6
Age group 0-14 years (%)	47.6
Age group 60+ years (women and men, %)	2.3/2.1
Life expectancy at birth (women and men, years)	69/65
Infant mortality rate (per 1,000 births)	39
Total fertility rate (births per woman)	6
Contraceptive use (% of currently married women)	52k
Urban population (%)	52
Urban population growth rate (% per annum)	4.4
Rural population growth rate (% per annum)	2.7
Foreign-born (1985,%)	6.8
Refugees	5700
Government education expenditure (% of GDP)	3l
Primary-secondary gross enrolment ratio (f and m, per 100)	76/90
Third-level students (per 100,000 population)	1737l
Newspaper circulation (per 1,000 population)	23
Television receivers (per 1,000 population)	60
Intentional homicides (1986, per 100,000 population)	5
Parliamentary seats (women and men, %)	8/92

Environment	1990/91
Threatened species	19
Forested area (%)	3.9
CO_2 emissions (10,000 Mt)	8124
Energy consumption (1,000 Mt coal equiv)	1100
Precipitation (mm)	234l
Average temperature (January and July, centigrade)	7.7/27.2l

a April 1994. b 1990. c 1992. d 1988. e 1991. f 1980. g 1983. h 1984.
i 1987. j Includes Monaco. k Source: UNICEF94. l 1989.

Tajikistan

Region Central Asia
Location (longitude, latitude) 38°35'N 68°48'E[a]
Currency rouble
Surface area (square kms) 143100
Largest city (pop. in thousands) Dushanbe (607)[a]
UN membership date 02-Mar-1992
Major language(s) Tajik, Uzbek, Russian

Economic indicators	1985	1994
Exchange rate (US$)	...	2144.00
GDP (million US$)	7301	7683[b]
GDP (per capita US$)	1599	1412[b]
Long-term rate of change in GDP (% per annum)	7.2	−3.5[b]
Economically active female population (%)	55.7[c]	57.8
Economically active male population (%)	76.4[c]	79.1
Telephone lines (per 100 inhabitants)	4.1[d]	4.8[b]

Social indicators	1990/95
Contraceptive use (% of currently married women)	21
Refugees	3000

Environment	1990/91
Threatened species	23

a Capital city. b 1991. c 1980. d 1987.

Thailand

Region	South-eastern Asia
Location (longitude, latitude)	13°76'N 100°53'E
Currency	baht
Population (1995 est., in thousands)	58265
Surface area (square kms)	513115
Population density (pop. per square km)	111
Sex ratio (females per 100 males)	102
Largest city (pop. in thousands)	Bangkok (7087)
UN membership date	15-Dec-1946
Major language(s)	Thai, Lao, Chinese

Economic indicators	1985	1994
Exchange rate (US$)	26.65	24.97
Consumer price index (1980=100)	129	193
Unemployment (%)	...	3.5[ab]
Balance of payments, current account (million US$)	−1537	−6604[c]
Tourist arrivals (in thousands)	4231[d]	5136[c]
GDP (million US$)	38900	98343[b]
GDP (per capita US$)	760	1775[b]
Long-term rate of change in GDP (% per annum)	4.7	7.9[b]
Gross fixed capital formation (% of GDP)	27.2	41.8[b]
Economically active female population (%)	74.1[e]	64.8
Economically active male population (%)	85.3[e]	84.6
Annual growth of econ. active female pop. (%)	2.6[f]	...
Annual growth of econ. active male pop. (%)	3.1[f]	...
Labour force in industry (%)	10.3[e]	14.0[f]
Labour force in agriculture (%)	70.8[e]	64.0[f]
Agricultural production index (1979-1981=100)	125[d]	129[c]
Food production index (1979-1981=100)	124[d]	123[c]
Commercial energy production (1,000 Mt coal equiv)	741[e]	20291[b]
Motor vehicles (per 1,000 population)	28	42[b]
Telephone lines (per 100 inhabitants)	1.7[g]	2.8[b]

Largest export industries	Major trading partners		1992
(% of exports)	(% of exports)		(% of imports)
...	...	Japan	29
...	...	United States	12
...	...	Singapore	7

Social indicators	1990/95
Growth rate of population (% per annum)	1.3
Age group 0-14 years (%)	29.2
Age group 60+ years (women and men, %)	3.9/3.1
Life expectancy at birth (women and men, years)	72/67
Infant mortality rate (per 1,000 births)	26
Total fertility rate (births per woman)	2
Contraceptive use (% of currently married women)	66[g]
Urban population (%)	25
Urban population growth rate (% per annum)	3.9
Rural population growth rate (% per annum)	0.4
Foreign-born (1985, %)	0.7
Refugees	63600
Government education expenditure (% of GDP)	3
Primary-secondary gross enrolment ratio (f and m, per 100)	59/60
Third-level students (per 100,000 population)	1734[h]
Newspaper circulation (per 1,000 population)	73
Television receivers (per 1,000 population)	114
Intentional homicides (1986, per 100,000 population)	25
Parliamentary seats (women and men, %)	4/96

Environment	1990/91
Threatened species	118
Forested area (%)	27.5
CO_2 emissions (10,000 Mt)	27537
Energy consumption (1,000 Mt coal equiv)	833
Precipitation (mm)	1492[g]
Average temperature (January and July, centigrade)	26.1/28.4[e]

a Labour force sample surveys. b 1991. c 1992. d 1988. e 1980. f 1990. g 1987. h 1989.

The former Yugoslav Republic of Macedonia

Region Southern Europe
Location (longitude, latitude) 41°59'N 21°28'E
Largest city (pop. in thousands) Skopje (393)[a]
UN membership date 08-Apr-1993

Economic indicators	1985	1994
Economically active female population (%)	47.9[b]	49.9
Economically active male population (%)	75.7[b]	71.4

Social indicators	1990/95
Refugees	32000

a Capital city. b 1980.

Togo

Region	Western Africa
Location (longitude, latitude)	6°11'N 1°22'E[a]
Currency	CFA franc
Population (1995 est., in thousands)	4138
Surface area (square kms)	56785
Population density (pop. per square km)	64
Sex ratio (females per 100 males)	102
Largest city (pop. in thousands)	Lome (513)[a]
UN membership date	20-Sep-1960
Major language(s)	French, Ewe, Kabre, Gurma, Moba, Tem

Economic indicators	1985	1994
Exchange rate (US$)	378.05	528.15
Consumer price index (1980=100)	138	144[b]
Balance of payments, current account (million US$)	−34	−105[c]
Tourist arrivals (in thousands)	104[d]	49[c]
GDP (million US$)	740	1847[b]
GDP (per capita US$)	244	507[b]
Long-term rate of change in GDP (% per annum)	2.0	4.2[b]
Gross fixed capital formation (% of GDP)	22.9	...
Economically active female population (%)	51.5[e]	45.3
Economically active male population (%)	88.9[e]	87.3
Annual growth of econ. active female pop. (%)	1.9[f]	...
Annual growth of econ. active male pop. (%)	2.5[f]	...
Labour force in industry (%)	39.1[e]	24.1[g]
Labour force in agriculture (%)	2.6[e]	8.9[g]
Agricultural production index (1979-1981=100)	127[d]	139[c]
Food production index (1979-1981=100)	123[d]	131[c]
Commercial energy production (1,000 Mt coal equiv)	0[e]	1[b]
Motor vehicles (per 1,000 population)	1	...
Telephone lines (per 100 inhabitants)	0.3[g]	0.3[b]

Social indicators	1990/95
Growth rate of population (% per annum)	3.2
Age group 0-14 years (%)	45.7
Age group 60+ years (women and men, %)	2.7/2.2
Life expectancy at birth (women and men, years)	57/53
Infant mortality rate (per 1,000 births)	85
Total fertility rate (births per woman)	7
Contraceptive use (% of currently married women)	12[d]
Urban population (%)	31
Urban population growth rate (% per annum)	4.8
Rural population growth rate (% per annum)	2.5
Foreign-born (1985,%)	3.5
Refugees	3400
Government education expenditure (% of GDP)	5
Primary-secondary gross enrolment ratio (f and m, per 100)	51/87
Third-level students (per 100,000 population)	226[h]
Newspaper circulation (per 1,000 population)	3
Television receivers (per 1,000 population)	6
Parliamentary seats (women and men, %)	6/94

Environment	1990/91
Threatened species	17
Forested area (%)	10.7
CO_2 emissions (10,000 Mt)	197
Energy consumption (1,000 Mt coal equiv)	78
Precipitation (mm)	893[g]
Average temperature (January and July, centigrade)	26.8/24.6[e]

a Capital city. b 1991. c 1992. d 1988. e 1980. f 1990. g 1987. h 1989.

Tonga

Region Oceania-Polynesia
Location (longitude, latitude) 21°08'S 175°12'W[a]
Currency pa'anga
Surface area (square kms) 747
Population density (pop. per square km) 126
Largest city (pop. in thousands) Nuku'alofa (34)[a]
Major language(s) Tongan, English

Economic indicators	1985	1994
Exchange rate (US$)	1.47	1.30
Consumer price index (1980=100)	117	234[b]
Balance of payments, current account (million US$)	−1	−1[c]
Tourist arrivals (in thousands)	19[d]	23[c]
GDP (million US$)	56	122[e]
GDP (per capita US$)	594	1262[e]
Long-term rate of change in GDP (% per annum)	5.4	3.7[e]
Annual growth of econ. active female pop. (%)	8.1[f]	...
Annual growth of econ. active male pop. (%)	0.3[f]	...
Agricultural production index (1979-1981=100)	66[d]	86[c]
Food production index (1979-1981=100)	66[d]	86[c]
Motor vehicles (per 1,000 population)	28	62[e]
Telephone lines (per 100 inhabitants)	2.9[h]	5.2[e]

Social indicators	1990/95
Infant mortality rate (per 1,000 births)	5[i]
Contraceptive use (% of currently married women)	74[j]
Urban population (%)	41
Urban population growth rate (% per annum)	3.7
Rural population growth rate (% per annum)	−1.4
Foreign-born (1985,%)	3.4
Government education expenditure (% of GDP)	4[k]
Newspaper circulation (per 1,000 population)	73
Intentional homicides (1986, per 100,000 population)	6
Parliamentary seats (women and men, %)	3/97

Environment	1990/91
Threatened species	10
CO_2 emissions (10,000 Mt)	20
Energy consumption (1,000 Mt coal equiv)	361

a Capital city. b December 1993. c 1992. d 1988. e 1991. f 1990.
g 1980. h 1987. i 1985. j Source: NATHR. k 1986.

Trinidad and Tobago

Region	Caribbean
Location (longitude, latitude)	10°66'N 61°52'W[a]
Currency	Trinidad and Tobago dollar
Population (1995 est., in thousands)	1305
Surface area (square kms)	5130
Population density (pop. per square km)	244
Sex ratio (females per 100 males)	102
Largest city (pop. in thousands)	Port-of-Spain (63)[a]
UN membership date	18-Sep-1962
Major language(s)	English

Economic indicators	1985	1994
Exchange rate (US$)	3.60	5.90
Consumer price index (1980=100)	181	385[b]
Balance of payments, current account (million US$)	−83	122[c]
Tourist arrivals (in thousands)	186[d]	235[c]
GDP (million US$)	7266	5275[e]
GDP (per capita US$)	6264	4217[e]
Long-term rate of change in GDP (% per annum)	−5.6	3.1[c]
Gross fixed capital formation (% of GDP)	20.9	12.7[f]
Economically active female population (%)	31.5[g]	34.0
Economically active male population (%)	79.4[g]	80.9
Annual growth of econ. active female pop. (%)	2.3[f]	...
Annual growth of econ. active male pop. (%)	2.3[f]	...
Labour force in industry (%)	36.9[g]	26.9[c]
Labour force in agriculture (%)	10.2[g]	11.6[c]
Agricultural production index (1979-1981=100)	91[d]	92[c]
Food production index (1979-1981=100)	94[d]	95[c]
Commercial energy production (1,000 Mt coal equiv)	20697[g]	18169[e]
Motor vehicles (per 1,000 population)	277	165[f]
Telephone lines (per 100 inhabitants)	13.3[h]	14.1[e]

Largest export industries		Major trading partners			1992
	(% of exports)		(% of exports)		(% of imports)
Chemicals	54	United States	49	United States	42
Mining, quarrying	26	Nether. Antilles	7	Venezuela	10
Basic metal industry	8	Barbados	3	United Kingdom	8

Social indicators	1990/95
Growth rate of population (% per annum)	1.1
Age group 0-14 years (%)	33.8
Age group 60+ years (women and men, %)	4.5/3.5
Life expectancy at birth (women and men, years)	74/69
Infant mortality rate (per 1,000 births)	18
Total fertility rate (births per woman)	3
Contraceptive use (% of currently married women)	53[h]
Urban population (%)	67
Urban population growth rate (% per annum)	1.6
Rural population growth rate (% per annum)	0.6
Foreign-born (1985,%)	5.4
Government education expenditure (% of GDP)	4
Primary-secondary gross enrolment ratio (f and m, per 100)	90/89
Third-level students (per 100,000 population)	563[i]
Newspaper circulation (per 1,000 population)	77
Television receivers (per 1,000 population)	315
Intentional homicides (1986, per 100,000 population)	8
Parliamentary seats (women and men, %)	14/86

Environment	1990/91
Threatened species	10
Forested area (%)	42.9
CO_2 emissions (10,000 Mt)	5030
Energy consumption (1,000 Mt coal equiv)	8063
Precipitation (mm)	1772[h]
Average temperature (January and July, centigrade)	24.3/25.9[g]

a Capital city. b June 1994. c 1992. d 1988. e 1991. f 1990. g 1980.
h 1987. i 1989.

Tunisia

Region	Northern Africa
Location (longitude, latitude)	36°79'N 10°19'E
Currency	Tunisian dinar
Population (1995 est., in thousands)	8933
Surface area (square kms)	163610
Population density (pop. per square km)	51
Sex ratio (females per 100 males)	98
Largest city (pop. in thousands)	Tunis (1755)
UN membership date	12-Nov-1956
Major language(s)	Arabic

Economic indicators	1985	1994
Exchange rate (US$)	0.76	0.98
Consumer price index (1980=100)	158	123[a]
Industrial production index (1980 = 100)	112	106[b]
Balance of payments, current account (million US$)	−587	−945[c]
Tourist arrivals (in thousands)	3468[d]	3540[c]
GDP (million US$)	8275	12759[e]
GDP (per capita US$)	1140	1551[e]
Long-term rate of change in GDP (% per annum)	5.7	3.1[e]
Gross fixed capital formation (% of GDP)	26.8	...
Economically active female population (%)	21.7[f]	25.9
Economically active male population (%)	79.4[f]	79.2
Annual growth of econ. active female pop. (%)	7.3[g]	...
Annual growth of econ. active male pop. (%)	2.8[g]	...
Agricultural production index (1979-1981=100)	101[d]	158[c]
Food production index (1979-1981=100)	101[d]	158[c]
Commercial energy production (1,000 Mt coal equiv)	8666[f]	7900[e]
Motor vehicles (per 1,000 population)	45	...
Telephone lines (per 100 inhabitants)	3.0[h]	4.1[e]

Largest export industries		Major trading partners			1992
	(% of exports)		(% of exports)		(% of imports)
Textiles	43	France[i]	27	France[i]	26
Chemicals	16	Italy	17	Italy	18
Mining, quarrying	14	Germany	17	Germany	14

Social indicators	1990/95
Growth rate of population (% per annum)	2.1
Age group 0-14 years (%)	35.2
Age group 60+ years (women and men, %)	3.5/3.4
Life expectancy at birth (women and men, years)	69/67
Infant mortality rate (per 1,000 births)	43
Total fertility rate (births per woman)	3
Contraceptive use (% of currently married women)	50[d]
Urban population (%)	59
Urban population growth rate (% per annum)	3.1
Rural population growth rate (% per annum)	0.7
Foreign-born (1985, %)	0.5
Refugees	100
Government education expenditure (% of GDP)	6
Primary-secondary gross enrolment ratio (f and m, per 100)	74/86
Third-level students (per 100,000 population)	838
Newspaper circulation (per 1,000 population)	37
Television receivers (per 1,000 population)	79
Parliamentary seats (women and men, %)	4/96

Environment	1990/91
Threatened species	20
Forested area (%)	4.0
CO_2 emissions (10,000 Mt)	4042
Energy consumption (1,000 Mt coal equiv)	839
Precipitation (mm)	466[h]
Average temperature (January and July, centigrade)	11.0/25.9[f]

a May 1994. b March 1994. c 1992. d 1988. e 1991. f 1980. g 1990. h 1987. i Includes Monaco.

Turkey

Region Western Asia
Location (longitude, latitude) 41°01'N 28°95'E
Currency Turkish lira
Population (1995 est., in thousands) 62032
Surface area (square kms) 779452
Population density (pop. per square km) 78
Sex ratio (females per 100 males) 96
Largest city (pop. in thousands) Istanbul (6507)
UN membership date 24-Oct-1945
Major language(s) Turkish, Kurdish

Economic indicators	1985	1994
Exchange rate (US$)	576.86	8564.40[a]
Consumer price index (1980=100)	283	4543[b]
Industrial production index (1980 = 100)	149	256
Balance of payments, current account (million US$)	−1013	−6380[c]
Tourist arrivals (in thousands)	3715[d]	6549[a]
GDP (million US$)	68432	150168[e]
GDP (per capita US$)	1359	2627[e]
Long-term rate of change in GDP (% per annum)	4.4	0.9[e]
Gross fixed capital formation (% of GDP)	21.8	...
Economically active female population (%)	45.6[f]	45.0
Economically active male population (%)	84.5[f]	83.0
Annual growth of econ. active female pop. (%)	1.7[g]	...
Annual growth of econ. active male pop. (%)	2.6[g]	...
Labour force in industry (%)	35.1[h]	23.0[a]
Labour force in agriculture (%)	4.3[h]	45.0[a]
Agricultural production index (1979-1981=100)	128[d]	131[a]
Food production index (1979-1981=100)	128[d]	132[a]
Commercial energy production (1,000 Mt coal equiv)	14085[f]	26315[e]
Motor vehicles (per 1,000 population)	27	42[e]
Telephone lines (per 100 inhabitants)	7.7[i]	14.3[e]

Largest export industries	Major trading partners		1992
(% of exports)	(% of exports)	(% of imports)	
Textiles 40	Germany 25	Germany 16	
Agriculture 15	Italy 6	United States 11	
Metal manufacture 11	United States 6	Italy 8	

Social indicators	1990/95
Growth rate of population (% per annum)	2.0
Age group 0-14 years (%)	33.7
Age group 60+ years (women and men, %)	4.0/3.7
Life expectancy at birth (women and men, years)	70/65
Infant mortality rate (per 1,000 births)	56
Total fertility rate (births per woman)	3
Contraceptive use (% of currently married women)	63[d]
Urban population (%)	69
Urban population growth rate (% per annum)	4.5
Rural population growth rate (% per annum)	−2.4
Foreign-born (1985,%)	1.9
Refugees	28500
Government education expenditure (% of GDP)	3
Primary-secondary gross enrolment ratio (f and m, per 100)	72/89
Third-level students (per 100,000 population)	1342
Newspaper circulation (per 1,000 population)	71
Television receivers (per 1,000 population)	175
Parliamentary seats (women and men, %)	2/98

Environment	1990/91
Threatened species	79
Forested area (%)	25.9
CO_2 emissions (10,000 Mt)	38907
Energy consumption (1,000 Mt coal equiv)	1036
Precipitation (mm)	669[j]
Average temperature (January and July, centigrade)	4.6/22.2[f]

a 1992. b August 1994. c 1993. d 1988. e 1991. f 1980. g 1990. h 1982. i 1987.

Turkmenistan

Region Central Asia
Location (longitude, latitude) 37A8'N 58'E[a]
Currency rouble
Surface area (square kms) 488100
Largest city (pop. in thousands) Ashkhabad (412)[a]
UN membership date 02-Mar-1992
Major language(s) Turkmen, Russian, Uzbek

Economic indicators	1985	1994
Exchange rate (US$)	...	2.00
GDP (million US$)	6648	8429[b]
GDP (per capita US$)	2058	2239[b]
Long-term rate of change in GDP (% per annum)	1.3	−1.1[b]
Economically active female population (%)	56.7[c]	59.2
Economically active male population (%)	77.9[c]	80.4
Telephone lines (per 100 inhabitants)	5.3[d]	6.3[b]

Social indicators	1990/95
Contraceptive use (% of currently married women)	20

Environment	1990/91
Threatened species	35

a Capital city. b 1991. c 1980. d 1987.

Uganda

Region	Eastern Africa
Location (longitude, latitude)	0°29'N 32°46'E
Currency	Uganda shilling
Population (1995 est., in thousands)	20405
Surface area (square kms)	235880
Population density (pop. per square km)	83
Sex ratio (females per 100 males)	101
Largest city (pop. in thousands)	Kampala (754)
UN membership date	25-Oct-1962
Major language(s)	English, Ganda, Nkore-Kiga

Economic indicators	1985	1994
Exchange rate (US$)	14.00	918.91[a]
Consumer price index (1980=100)	499	288[b]
Balance of payments, current account (million US$)	5	−100[c]
Tourist arrivals (in thousands)	40[d]	50[c]
GDP (million US$)	3200	3051[e]
GDP (per capita US$)	213	168[e]
Long-term rate of change in GDP (% per annum)	2.0	4.1[e]
Economically active female population (%)	66.6[f]	59.5
Economically active male population (%)	92.8[f]	91.5
Annual growth of econ. active female pop. (%)	2.5[g]	...
Annual growth of econ. active male pop. (%)	3.0[g]	...
Agricultural production index (1979-1981=100)	126[d]	147[c]
Food production index (1979-1981=100)	126[d]	148[c]
Commercial energy production (1,000 Mt coal equiv)	80[f]	95[e]
Motor vehicles (per 1,000 population)	1	2[e]
Telephone lines (per 100 inhabitants)	0.2[h]	0.2[e]

Social indicators	1990/95
Growth rate of population (% per annum)	3.0
Age group 0-14 years (%)	48.7
Age group 60+ years (women and men, %)	2.1/1.8
Life expectancy at birth (women and men, years)	43/41
Infant mortality rate (per 1,000 births)	104
Total fertility rate (births per woman)	7
Contraceptive use (% of currently married women)	5[i]
Urban population (%)	13
Urban population growth rate (% per annum)	5.3
Rural population growth rate (% per annum)	2.7
Foreign-born (1985,%)	1.3
Refugees	196300
Government education expenditure (% of GDP)	3[h]
Primary-secondary gross enrolment ratio (f and m, per 100)	41/52[i]
Third-level students (per 100,000 population)	82[i]
Newspaper circulation (per 1,000 population)	2
Television receivers (per 1,000 population)	10
Parliamentary seats (women and men, %)	13/87

Environment	1990/91
Threatened species	37
Forested area (%)	23.6
CO_2 emissions (10,000 Mt)	249
Energy consumption (1,000 Mt coal equiv)	29

a August 1994. b June 1994. c 1992. d 1988. e 1991. f 1980. g 1990. h 1987. i 1989. j 1986.

Ukraine

Region Eastern Europe
Location (longitude, latitude) 50°44'N 30°53'E
Currency rouble
Surface area (square kms) 603700
Largest city (pop. in thousands) Kiev (2638)
UN membership date 24-Oct-1945
Major language(s) Ukrainian, Russian

Economic indicators	1985	1994
Exchange rate (US$)	...	45.00
Consumer price index (1980=100)	104	...
Industrial production index (1980 = 100)	119	...
GDP (million US$)	150941	166972[a]
GDP (per capita US$)	2963	3208[a]
Long-term rate of change in GDP (% per annum)	1.0	−5.8[a]
Economically active female population (%)	54.7[b]	52.3
Economically active male population (%)	76.7[b]	75.5
Motor vehicles (per 1,000 population)	48	71[a]
Telephone lines (per 100 inhabitants)	12.9[c]	15.6

Social indicators	1990/95
Contraceptive use (% of currently married women)	23
Newspaper circulation (per 1,000 population)	374
Television receivers (per 1,000 population)	487
Intentional homicides (1986, per 100,000 population)	8[d]

Environment	1990/91
Threatened species	51

a 1991. b 1980. c 1987. d Source: DYB92.

United Arab Emirates

Region	Western Asia
Location (longitude, latitude)	24°28'N 54°22'E[a]
Currency	UAE dirham
Population (1995 est., in thousands)	1785
Surface area (square kms)	83600
Population density (pop. per square km)	19
Sex ratio (females per 100 males)	52
Largest city (pop. in thousands)	Abu Dhabi (589)[a]
UN membership date	09-Dec-1971
Major language(s)	Arabic, Malayalam, Persian, Pashto

Economic indicators	1985	1994
Exchange rate (US$)	3.67	3.67
Tourist arrivals (in thousands)	479[b]	400[c]
GDP (million US$)	27081	34323[d]
GDP (per capita US$)	20075	21057[d]
Long-term rate of change in GDP (% per annum)	−2.4	2.0[d]
Gross fixed capital formation (% of GDP)	24.6	20.0[d]
Economically active female population (%)	15.9[e]	20.8
Economically active male population (%)	94.1[e]	92.0
Annual growth of econ. active female pop. (%)	13.2[f]	...
Annual growth of econ. active male pop. (%)	11.0[f]	...
Commercial energy production (1,000 Mt coal equiv)	128789[e]	203604[d]
Telephone lines (per 100 inhabitants)	18.0[g]	29.5[d]

Social indicators	1990/95
Growth rate of population (% per annum)	2.3
Age group 0-14 years (%)	28.9
Age group 60+ years (women and men, %)	1.2/2.9
Life expectancy at birth (women and men, years)	74/70
Infant mortality rate (per 1,000 births)	22
Total fertility rate (births per woman)	5
Urban population (%)	84
Urban population growth rate (% per annum)	3.1
Rural population growth rate (% per annum)	−1.2
Foreign-born (1985,%)	67.6
Government education expenditure (% of GDP)	2
Primary-secondary gross enrolment ratio (f and m, per 100)	95/92
Third-level students (per 100,000 population)	642
Newspaper circulation (per 1,000 population)	157
Television receivers (per 1,000 population)	107
Intentional homicides (1986, per 100,000 population)	3
Parliamentary seats (women and men, %)	0/100

Environment	1990/91
Threatened species	9
CO_2 emissions (10,000 Mt)	16228
Energy consumption (1,000 Mt coal equiv)	6415

a Capital city. b 1988. c 1992. d 1991. e 1980. f 1990. g 1987.

United Kingdom

Region	Northern Europe
Location (longitude, latitude)	51°01'N 0°13'W
Currency	pounds sterling
Population (1995 est., in thousands)	58093
Surface area (square kms)	244100
Population density (pop. per square km)	235
Sex ratio (females per 100 males)	104
Largest city (pop. in thousands)	London (7335)
UN membership date	24-Oct-1945
Major language(s)	English

Economic indicators	1985	1994
Exchange rate (US$)	0.69	0.63
Consumer price index (1980=100)	142	217
Industrial production index (1980 = 100)	108	126[a]
Unemployment (%)	...	9.4
Balance of payments, current account (million US$)	3309	−16064[b]
Tourist arrivals (in thousands)	15799[c]	18535[d]
GDP (million US$)	461863	978191[e]
GDP (per capita US$)	8158	16994[e]
Long-term rate of change in GDP (% per annum)	3.8	−2.3[e]
Gross fixed capital formation (% of GDP)	16.9	16.7[e]
Economically active female population (%)	45.2[f]	46.1
Economically active male population (%)	77.0[f]	77.3
Annual growth of econ. active female pop. (%)	0.8[g]	...
Annual growth of econ. active male pop. (%)	0.2[g]	...
Labour force in industry (%)	37.2[f]	26.2[d]
Labour force in agriculture (%)	2.6[f]	2.2[d]
Agricultural production index (1979-1981=100)	107[c]	113[d]
Food production index (1979-1981=100)	106[c]	113[d]
Commercial energy production (1,000 Mt coal equiv)	284951[f]	299244[e]
Motor vehicles (per 1,000 population)	348	415[e]
Telephone lines (per 100 inhabitants)	39.8[h]	44.5[e]

Largest export industries		Major trading partners			1992
	(% of exports)		(% of exports)		(% of imports)
Metal manufacture	48	Germany	14	Germany	15
Chemicals	19	United States	11	United States	11
Food, beverages, tobacco	7	France[i]	11	France[i]	10

Social indicators	1990/95
Growth rate of population (% per annum)	0.2
Age group 0-14 years (%)	19.7
Age group 60+ years (women and men, %)	11.7/8.7
Life expectancy at birth (women and men, years)	79/74
Infant mortality rate (per 1,000 births)	7
Total fertility rate (births per woman)	2
Contraceptive use (% of currently married women)	81
Urban population (%)	89
Urban population growth rate (% per annum)	0.3
Rural population growth rate (% per annum)	−0.5
Foreign-born (1985,%)	6.5
Refugees	100000
Government education expenditure (% of GDP)	5
Primary-secondary gross enrolment ratio (f and m, per 100)	95/93[j]
Third-level students (per 100,000 population)	2063[j]
Newspaper circulation (per 1,000 population)	394[c]
Television receivers (per 1,000 population)	434
Intentional homicides (1986, per 100,000 population)	1[k]
Parliamentary seats (women and men, %)	9/91

Environment	1990/91
Threatened species	29
Forested area (%)	9.8
CO_2 emissions (10,000 Mt)	157521
Energy consumption (1,000 Mt coal equiv)	22379
Precipitation (mm)	778[h]
Average temperature (January and July, centigrade)	4.2/17.6[f]

a June 1994. b 1993. c 1988. d 1992. e 1991. f 1980. g 1990. h 1987.
i Includes Monaco. j 1989. k England and Wales only.

United Republic of Tanzania

Region	Eastern Africa
Location (longitude, latitude)	6°81'S 39°29'E
Currency	Tanzanian shilling
Population (1995 est., in thousands)	30742
Surface area (square kms)	945087
Population density (pop. per square km)	30
Sex ratio (females per 100 males)	102
Largest city (pop. in thousands)	Dar Es Salaam (1436)
UN membership date	14-Dec-1961
Major language(s)	Swahili, English, Nyamwezi

Economic indicators	1985	1994
Exchange rate (US$)	16.50	519.49[a]
Consumer price index (1980=100)	374	2547[b]
Balance of payments, current account (million US$)	−375	−426[c]
Tourist arrivals (in thousands)	130[d]	202[e]
GDP (million US$)	6904	3150[f]
GDP (per capita US$)	315	117[f]
Long-term rate of change in GDP (% per annum)	2.6	3.8[f]
Gross fixed capital formation (% of GDP)	14.0	38.0[f]
Economically active female population (%)	84.7[g]	74.8
Economically active male population (%)	89.7[g]	88.3
Annual growth of econ. active female pop. (%)	2.6[c]	...
Annual growth of econ. active male pop. (%)	3.2[c]	...
Labour force in industry (%)	25.9[g]	25.6[h]
Labour force in agriculture (%)	16.6[g]	13.9[h]
Agricultural production index (1979-1981=100)	115[d]	117[e]
Food production index (1979-1981=100)	116[d]	118[e]
Commercial energy production (1,000 Mt coal equiv)	66[g]	80[f]
Motor vehicles (per 1,000 population)	4	4[c]
Telephone lines (per 100 inhabitants)	0.3[i]	0.3[f]

Social indicators	1990/95
Growth rate of population (% per annum)	3.4
Age group 0-14 years (%)	48.0
Age group 60+ years (women and men, %)	2.2/1.8
Life expectancy at birth (women and men, years)	52/49
Infant mortality rate (per 1,000 births)	102
Total fertility rate (births per woman)	7
Contraceptive use (% of currently married women)	10
Urban population (%)	24
Urban population growth rate (% per annum)	6.5
Rural population growth rate (% per annum)	2.4
Foreign-born (1985, %)	2.7
Refugees	292100
Government education expenditure (% of GDP)	5
Primary-secondary gross enrolment ratio (f and m, per 100)	39/41[j]
Third-level students (per 100,000 population)	25[i]
Newspaper circulation (per 1,000 population)	8
Television receivers (per 1,000 population)	2
Parliamentary seats (women and men, %)	11/89

Environment	1990/91
Threatened species	64
Forested area (%)	43.3
CO_2 emissions (10,000 Mt)	589
Energy consumption (1,000 Mt coal equiv)	5373
Precipitation (mm)	1043[i]
Average temperature (January and July, centigrade)	27.3/23.3[g]

a June 1994. b December 1993. c 1990. d 1988. e 1992. f 1991. g 1980. h 1984. i 1987. j 1989.

United States of America

Region	Northern America
Location (longitude, latitude)	40°76'N 73°96'W
Currency	US dollar
Population (1995 est., in thousands)	263138
Surface area (square kms)	9809431
Population density (pop. per square km)	26
Sex ratio (females per 100 males)	105
Largest city (pop. in thousands)	New York (16056)
UN membership date	24-Oct-1945
Major language(s)	English

Economic indicators	1985	1994
Exchange rate (US$)	1.00	1.00
Consumer price index (1980=100)	131	181
Industrial production index (1980 = 100)	112	142
Unemployment (%)	...	5.3[a]
Balance of payments, current account (million US$)	−122	−109[b]
Tourist arrivals (in thousands)	34095[c]	44647[d]
GDP (million US$)	4016649	5610800[e]
GDP (per capita US$)	16844	22219[e]
Long-term rate of change in GDP (% per annum)	3.2	−1.2[e]
Gross fixed capital formation (% of GDP)	19.5	15.4[e]
Economically active female population (%)	49.8[f]	50.1
Economically active male population (%)	75.7[f]	76.5
Annual growth of econ. active female pop. (%)	2.4[g]	...
Annual growth of econ. active male pop. (%)	1.3[g]	...
Labour force in industry (%)	30.8[f]	24.9[d]
Labour force in agriculture (%)	3.6[f]	2.9[d]
Agricultural production index (1979-1981=100)	94[c]	114[d]
Food production index (1979-1981=100)	94[c]	114[d]
Commercial energy production (1,000 Mt coal equiv)	2102772[f]	2317467[e]
Motor vehicles (per 1,000 population)	717	755[g]
Telephone lines (per 100 inhabitants)	52.2[h]	55.3[e]

Largest export industries		Major trading partners			1992
	(% of exports)		(% of exports)		(% of imports)
Metal manufacture	57	Canada	20	Japan	18
Chemicals	14	Japan	11	Canada	18
Agriculture	7	Mexico	9	Mexico	7

Social indicators	1990/95
Growth rate of population (% per annum)	1.0
Age group 0-14 years (%)	21.9
Age group 60+ years (women and men, %)	9.5/6.9
Life expectancy at birth (women and men, years)	79/73
Infant mortality rate (per 1,000 births)	8
Total fertility rate (births per woman)	2
Contraceptive use (% of currently married women)	74[c]
Urban population (%)	76
Urban population growth rate (% per annum)	1.3
Rural population growth rate (% per annum)	0.2
Foreign-born (1985,%)	7.0
Refugees	473000
Government education expenditure (% of GDP)	5[i]
Primary-secondary gross enrolment ratio (f and m, per 100)	99/99[i]
Third-level students (per 100,000 population)	5596[i]
Newspaper circulation (per 1,000 population)	249
Television receivers (per 1,000 population)	814
Intentional homicides (1986, per 100,000 population)	9
Parliamentary seats (women and men, %)	11/89

Environment	1990/91
Threatened species	961
Forested area (%)	31.3
CO_2 emissions (10,000 Mt)	1345969
Energy consumption (1,000 Mt coal equiv)	38
Precipitation (mm)	1123[h]
Average temperature (January and July, centigrade)	0.9/24.9[f]

a Labour force sample surveys. b 1993. c 1988. d 1992. e 1991. f 1980. g 1990. h 1987. i 1989.

Uruguay

Region	South America
Location (longitude, latitude)	34°91'S 56°19'W
Currency	Uruguayan new peso
Population (1995 est., in thousands)	3186
Surface area (square kms)	177414
Population density (pop. per square km)	18
Sex ratio (females per 100 males)	105
Largest city (pop. in thousands)	Montevideo (1287)
UN membership date	18-Dec-1945
Major language(s)	Spanish

Economic indicators	1985	1994
Exchange rate (US$)	125.00[a]	5.59
Consumer price index (1980=100)	637	89684[b]
Industrial production index (1980 = 100)	77	109[c]
Balance of payments, current account (million US$)	−99	−207[d]
Tourist arrivals (in thousands)	1036[e]	1802[d]
GDP (million US$)	4719	9479[f]
GDP (per capita US$)	1569	3046[f]
Long-term rate of change in GDP (% per annum)	1.5	1.9[f]
Gross fixed capital formation (% of GDP)	9.6	25.0[f]
Economically active female population (%)	30.8[g]	32.2
Economically active male population (%)	76.4[g]	73.6
Annual growth of econ. active female pop. (%)	1.4[c]	...
Annual growth of econ. active male pop. (%)	0.2[c]	...
Labour force in industry (%)	30.2[h]	29.5[d]
Labour force in agriculture (%)	4.5[d]	...
Agricultural production index (1979-1981=100)	116[e]	123[d]
Food production index (1979-1981=100)	115[e]	121[d]
Commercial energy production (1,000 Mt coal equiv)	279[e]	751[f]
Motor vehicles (per 1,000 population)	117	142[f]
Telephone lines (per 100 inhabitants)	10.3[i]	14.5[f]

Largest export industries		Major trading partners			1992
	(% of exports)		(% of exports)		(% of imports)
Agriculture	15	Brazil	18	Brazil	24
Chemicals	8	Argentina	15	Argentina	17
Metal manufacture	4	United States	11	United States	9

Social indicators	1990/95
Growth rate of population (% per annum)	0.6
Age group 0-14 years (%)	24.4
Age group 60+ years (women and men, %)	9.7/7.3
Life expectancy at birth (women and men, years)	76/69
Infant mortality rate (per 1,000 births)	20
Total fertility rate (births per woman)	2
Urban population (%)	90
Urban population growth rate (% per annum)	0.9
Rural population growth rate (% per annum)	−2.1
Foreign-born (1985,%)	3.5
Refugees	100
Government education expenditure (% of GDP)	3
Primary-secondary gross enrolment ratio (f and m, per 100)	85/85[g]
Third-level students (per 100,000 population)	3751[e]
Newspaper circulation (per 1,000 population)	233
Television receivers (per 1,000 population)	231
Intentional homicides (1986, per 100,000 population)	5
Parliamentary seats (women and men, %)	6/94

Environment	1990/91
Threatened species	22
Forested area (%)	3.8
CO_2 emissions (10,000 Mt)	1217
Energy consumption (1,000 Mt coal equiv)	10921
Precipitation (mm)	1014[i]
Average temperature (January and July, centigrade)	22.5/10.5

a Selling rate. b August 1994. c 1990. d 1992. e 1988. f 1991. g 1980.
h 1984. i 1987.

US Virgin Islands

Region	Caribbean
Location (longitude, latitude)	18°21'N 64°56'W[a]
Currency	US dollar
Population (1995 est., in thousands)	97
Surface area (square kms)	342
Population density (pop. per square km)	345
Sex ratio (females per 100 males)	109
Largest city (pop. in thousands)	Charlotte Amalie (13)[a]
Major language(s)	English

Economic indicators	1985	1994
Consumer price index (1980=100)	152[b]	167
Unemployment (%)	...	5.4
Tourist arrivals (in thousands)	556[c]	487[d]
Labour force in industry (%)	17.9[e]	12.3[f]
Labour force in agriculture (%)	0.5[e]	0.5[f]
Telephone lines (per 100 inhabitants)	38.9[g]	47.8[f]

Social indicators	1990/95
Age group 0-14 years (%)	36.0[e]
Age group 60+ years (women and men, %)	3.8/3.3[e]
Infant mortality rate (per 1,000 births)	19
Urban population (%)	49
Urban population growth rate (% per annum)	1.1
Rural population growth rate (% per annum)	−0.7
Foreign-born (1985,%)	56.6
Newspaper circulation (per 1,000 population)	179
Television receivers (per 1,000 population)	607

Environment	1990/91
Threatened species	6
Forested area (%)	5.9
CO_2 emissions (10,000 Mt)	2146
Energy consumption (1,000 Mt coal equiv)	766

a Capital city. b 1990. c 1988. d 1992. e 1980. f 1991. g 1987.

Uzbekistan

Region Central Asia
Location (longitude, latitude) 41°16'N 69°16'E
Currency rouble
Surface area (square kms) 447400
Largest city (pop. in thousands) Tashkent (2109)
UN membership date 02-Mar-1992
Major language(s) Uzbek, Russian

Economic indicators	1985	1994
Exchange rate (US$)	...	1740.00[a]
GDP (million US$)	31945	35264[b]
GDP (per capita US$)	1758	1681[b]
Long-term rate of change in GDP (% per annum)	4.1	−10.2[b]
Economically active female population (%)	57.9[c]	61.2
Economically active male population (%)	75.4[c]	78.7
Telephone lines (per 100 inhabitants)	5.8[d]	7.1[b]

Social indicators	1990/95
Contraceptive use (% of currently married women)	28

Environment	1990/91
Threatened species	27

a June 1994. b 1991. c 1980. d 1987.

Vanuatu

Region Oceania-Melanesia
Location (longitude, latitude) 17°45'S 168°18'E[a]
Currency vatu
Population (1995 est., in thousands) 150
Surface area (square kms) 12189
Population density (pop. per square km) 13
Sex ratio (females per 100 males) 92
Largest city (pop. in thousands) Villa (18)[a]
UN membership date 15-Sep-1981
Major language(s) English, French, Bislama

Economic indicators	1985	1994
Exchange rate (US$)	100.25	115.85
Consumer price index (1980=100)	115	117
Balance of payments, current account (million US$)	1	4[b]
Tourist arrivals (in thousands)	16[c]	43[b]
GDP (million US$)	118	179[d]
GDP (per capita US$)	889	1171[d]
Long-term rate of change in GDP (% per annum)	1.1	5.1[d]
Gross fixed capital formation (% of GDP)	22.7	40.8[e]
Economically active female population (%)	55.0[f]	50.9
Economically active male population (%)	85.6[f]	85.3
Agricultural production index (1979-1981=100)	105[c]	112[b]
Food production index (1979-1981=100)	106[c]	113[b]
Motor vehicles (per 1,000 population)	25	44[d]
Telephone lines (per 100 inhabitants)	1.6[g]	2.0[d]

Social indicators	1990/95
Growth rate of population (% per annum)	1.3
Age group 0-14 years (%)	45.6
Age group 60+ years (women and men, %)	2.1/2.2[hi]
Contraceptive use (% of currently married women)	15[j]
Urban population (%)	19
Urban population growth rate (% per annum)	3.3
Rural population growth rate (% per annum)	2.2
Foreign-born (1985,%)	3.1
Government education expenditure (% of GDP)	4
Television receivers (per 1,000 population)	9

Environment	1990/91
Threatened species	15
Forested area (%)	75.0
CO_2 emissions (10,000 Mt)	18
Energy consumption (1,000 Mt coal equiv)	209

a Capital city. b 1992. c 1988. d 1991. e 1990. f 1980. g 1987. h Source DYB90. i 1989. j Source: NATHR.

Venezuela

Region	South America
Location (longitude, latitude)	10°51'N 66°91'W
Currency	bolivare
Population (1995 est., in thousands)	21483
Surface area (square kms)	912050
Population density (pop. per square km)	22
Sex ratio (females per 100 males)	99
Largest city (pop. in thousands)	Caracas (2775)
UN membership date	15-Nov-1945
Major language(s)	Spanish

Economic indicators	1985	1994
Exchange rate (US$)	7.50	79.45[a]
Consumer price index (1980=100)	111	2264[b]
Industrial production index (1980 = 100)	181	...
Unemployment (%)	...	7.8[ac]
Balance of payments, current account (million US$)	3327	−3365[a]
Tourist arrivals (in thousands)	373[d]	434[a]
GDP (million US$)	61965	53440[a]
GDP (per capita US$)	3610	2705[a]
Long-term rate of change in GDP (% per annum)	0.2	10.4[e]
Gross fixed capital formation (% of GDP)	17.3	17.8[e]
Economically active female population (%)	28.6[f]	31.7
Economically active male population (%)	81.1[f]	80.6
Annual growth of econ. active female pop. (%)	5.7[g]	...
Annual growth of econ. active male pop. (%)	3.7[g]	...
Labour force in industry (%)	27.7[f]	24.6[g]
Labour force in agriculture (%)	15.0[f]	13.1[g]
Agricultural production index (1979-1981=100)	130[d]	136[a]
Food production index (1979-1981=100)	131[d]	138[a]
Commercial energy production (1,000 Mt coal equiv)	190996[f]	226297[e]
Motor vehicles (per 1,000 population)	117	106[g]
Telephone lines (per 100 inhabitants)	7.8[h]	8.1[e]

Largest export industries		Major trading partners			1992
	(% of exports)		(% of exports)		(% of imports)
Basic metal industry	43	United States	55	United States	48
Chemicals	16	Nether. Antilles	10	Japan	7
Metal manufacture	14	Netherlands	4	Germany	6

Social indicators	1990/95
Growth rate of population (% per annum)	2.1
Age group 0-14 years (%)	34.7
Age group 60+ years (women and men, %)	3.4/2.9
Life expectancy at birth (women and men, years)	74/67
Infant mortality rate (per 1,000 births)	33
Total fertility rate (births per woman)	3
Urban population (%)	93
Urban population growth rate (% per annum)	2.7
Rural population growth rate (% per annum)	−3.7
Foreign-born (1985, %)	6.1
Refugees	2000
Government education expenditure (% of GDP)	5[d]
Primary-secondary gross enrolment ratio (f and m, per 100)	90/86
Third-level students (per 100,000 population)	2670[d]
Newspaper circulation (per 1,000 population)	145
Television receivers (per 1,000 population)	162
Intentional homicides (1986, per 100,000 population)	13
Parliamentary seats (women and men, %)	10/90

Environment	1990/91
Threatened species	51
Forested area (%)	33.1
CO_2 emissions (10,000 Mt)	33189
Energy consumption (1,000 Mt coal equiv)	3490
Precipitation (mm)	854[h]
Average temperature (January and July, centigrade)	19.6/21.9[f]

a 1992. b August 1994. c Labour force sample surveys. d 1988. e 1991. f 1980. g 1990. h 1987.

Viet Nam

Region South-eastern Asia
Location (longitude, latitude) 10°45'N 106°40'E
Currency dong
Population (1995 est., in thousands) 73811
Surface area (square kms) 331689
Population density (pop. per square km) 206
Sex ratio (females per 100 males) 103
Largest city (pop. in thousands) Ho Chi Minh (3237)
UN membership date 20-Sep-1977
Major language(s) Vietnamese

Economic indicators	1985	1994
Exchange rate (US$)	22.50	11000.00
Tourist arrivals (in thousands)	148[a]	180[b]
GDP (million US$)	6990	9410[c]
GDP (per capita US$)	117	138[c]
Long-term rate of change in GDP (% per annum)	6.2	6.0[c]
Economically active female population (%)	68.2[d]	69.4
Economically active male population (%)	85.7[d]	85.1
Annual growth of econ. active female pop. (%)	2.6[e]	...
Annual growth of econ. active male pop. (%)	2.8[e]	...
Agricultural production index (1979-1981=100)	138[a]	166[b]
Food production index (1979-1981=100)	137[a]	165[b]
Commercial energy production (1,000 Mt coal equiv)	5582[d]	11302[c]
Telephone lines (per 100 inhabitants)	0.1[e]	0.1[c]

Social indicators	1990/95
Growth rate of population (% per annum)	2.0
Age group 0-14 years (%)	37.0
Age group 60+ years (women and men, %)	4.2/3.2
Life expectancy at birth (women and men, years)	66/62
Infant mortality rate (per 1,000 births)	36
Total fertility rate (births per woman)	4
Contraceptive use (% of currently married women)	53[a]
Urban population (%)	21
Urban population growth rate (% per annum)	2.9
Rural population growth rate (% per annum)	1.8
Foreign-born (1985,%)	0.5
Refugees	16300
Primary-secondary gross enrolment ratio (f and m, per 100)	67/71[f]
Third-level students (per 100,000 population)	214[d]
Newspaper circulation (per 1,000 population)	9
Television receivers (per 1,000 population)	41
Parliamentary seats (women and men, %)	18/82

Environment	1990/91
Threatened species	99
Forested area (%)	29.7
CO_2 emissions (10,000 Mt)	5615
Energy consumption (1,000 Mt coal equiv)	125
Precipitation (mm)	1808[g]
Average temperature (January and July, centigrade)	25.8/27.1[d]

a 1988. b 1992. c 1991. d 1980. e 1990. f 1985. g 1987.

Western Sahara

Region Northern Africa
Location (longitude, latitude) 27°10'N 13°11'W[a]
Surface area (square kms) 266000
Population density (pop. per square km) 1
Largest city (pop. in thousands) Elaaium (98)[a]

Social indicators	1990/95
Urban population (%)	60
Urban population growth rate (% per annum)	5.3
Rural population growth rate (% per annum)	2.5
Foreign-born (1985,%)	1.4
Television receivers (per 1,000 population)	21

Environment	1990/91
Threatened species	10
CO_2 emissions (10,000 Mt)	54
Energy consumption (1,000 Mt coal equiv)	388

a Capital city. b 1980. c 1991.

Yemen

Region	Western Asia
Location (longitude, latitude)	15°24'N 44°14'E[a]
Currency	Yemeni rial
Population (1995 est., in thousands)	13897
Sex ratio (females per 100 males)	102
Largest city (pop. in thousands)	Sanaa (500)[a]
Major language(s)	Arabic

Economic indicators	1985	1994
Exchange rate (US$)	12.01[bc]	25.00
Consumer price index (1980=100)	134[b]	...
Balance of payments, current account (million US$)	−287[b]	...
Tourist arrivals (in thousands)	60[bd]	72[be]
GDP (million US$)[b]	4181	7964[f]
GDP (per capita US$)[b]	549	658[f]
Long-term rate of change in GDP (% per annum)[b]	22.7	−3.9[f]
Gross fixed capital formation (% of GDP)	15.5[c]	...
Economically active female population (%)	8.4[g]	11.1
Economically active male population (%)	84.4[g]	84.1
Annual growth of econ. active female pop. (%)[b]	5.2[c]	...
Annual growth of econ. active male pop. (%)[b]	1.7[c]	...
Agricultural production index (1979-1981=100)	108[d]	111[e]
Food production index (1979-1981=100)	108[d]	110[e]
Commercial energy production (1,000 Mt coal equiv)	13348[f]	...
Motor vehicles (per 1,000 population)	6	...
Telephone lines (per 100 inhabitants)	0.8[h]	1.1[f]

Social indicators	1990/95
Growth rate of population (% per annum)	3.5
Age group 0-14 years (%)	49.3
Age group 60+ years (women and men, %)	2.1/1.7
Life expectancy at birth (women and men, years)	53/52
Infant mortality rate (per 1,000 births)	106
Total fertility rate (births per woman)	7
Contraceptive use (% of currently married women)[b]	7
Urban population (%)	34
Urban population growth rate (% per annum)	6.5
Rural population growth rate (% per annum)	2.1
Foreign-born (1985, %)	0.2
Refugees	59700
Government education expenditure (% of GDP)[b]	4[i]
Primary-secondary gross enrolment ratio (f and m, per 100)[b]	25/85
Third-level students (per 100,000 population)	274[d]
Newspaper circulation (per 1,000 population)[b]	13
Television receivers (per 1,000 population)	27
Parliamentary seats (women and men, %)	1/99

Environment	1990/91
Threatened species	18
Forested area (%)	7.7
CO_2 emissions (10,000 Mt)	2713
Energy consumption (1,000 Mt coal equiv)	328
Precipitation (mm)	39[j]
Average temperature (January and July, centigrade)	25.5/32.2[g]

a Capital city. b Data refer to Yemen (former, not including Democratic Yemen). c 1990. d 1988. e 1992. f 1991. g 1980. h 1987. i 1986. j Aden.

Yugoslavia

Region	Southern Europe
Location (longitude, latitude)	44°50'N 20°30'E[a]
Currency	Yugoslav new dinar
Population (1995 est., in thousands)	24113[b]
Sex ratio (females per 100 males)	102[b]
Largest city (pop. in thousands)	Belgrade (1575)
UN membership date	24-Oct-1945
Major language(s)	Serbo-Croatian, Slovenian, Albanian, Macedonian

Economic indicators	1985	1994
Exchange rate (US$)	5.73[bc]	1.55
Consumer price index (1980=100)	7[b]	74
Industrial production index (1980 = 100)	114[b]	78[d]
Unemployment (%)	...	19.5[bc]
Balance of payments, current account (million US$)	833	−1161[c]
Tourist arrivals (in thousands)[b]	9018[e]	700[c]
GDP (million US$)	17448	54826[c]
GDP (per capita US$)	1709	5182[c]
Long-term rate of change in GDP (% per annum)	0.3	−29.9
Gross fixed capital formation (% of GDP)[b]	21.8	14.7[f]
Economically active female population (%)	44.4[g]	49.2
Economically active male population (%)	75.1[g]	70.3
Annual growth of econ. active female pop. (%)[b]	1.2[f]	...
Annual growth of econ. active male pop. (%)[b]	0.7[f]	...
Labour force in industry (%)	51.6[bg]	51.2[bh]
Labour force in agriculture (%)	4.9[bg]	5.0[bh]
Agricultural production index (1979-1981=100)[b]	100[e]	82[d]
Food production index (1979-1981=100)[b]	100[e]	82[d]
Commercial energy production (1,000 Mt coal equiv)[b]	26605[g]	21363[c]
Motor vehicles (per 1,000 population)[b]	51	65[f]
Telephone lines (per 100 inhabitants)[b]	12.9[i]	10.5[c]

Social indicators	1990/95
Growth rate of population (% per annum)[b]	0.3
Age group 0-14 years (%)[b]	21.6
Age group 60+ years (women and men, %)[b]	9.3/6.9
Life expectancy at birth (women and men, years)[b]	75/69
Infant mortality rate (per 1,000 births)[b]	23
Total fertility rate (births per woman)[b]	2
Urban population (%)[b]	61
Urban population growth rate (% per annum)[b]	1.9
Rural population growth rate (% per annum)[b]	−2.0
Foreign-born (1985,%)[b]	1.6
Refugees[b]	516500
Government education expenditure (% of GDP)[b]	6
Primary-secondary gross enrolment ratio (f and m, per 100)[b]	84/85
Third-level students (per 100,000 population)[b]	1374
Newspaper circulation (per 1,000 population)[b]	96
Television receivers (per 1,000 population)[b]	195[j]
Intentional homicides (1986, per 100,000 population)	6[b]
Parliamentary seats (women and men, %)[b]	18/82[i]

Environment	1990/91
Threatened species[b]	62
Forested area (%)[b]	36.7
CO_2 emissions (10,000 Mt)[b]	23806
Energy consumption (1,000 Mt coal equiv)[b]	1731
Precipitation (mm)	701[i]
Average temperature (January and July, centigrade)	−0.2/22.6[g]

a Capital city. b Data refer to SFR Yugoslavia. c 1991. d 1992. e 1988. f 1990. g 1980. h 1989. i 1987. j 1985.

Zaire

Region	Middle Africa
Location (longitude, latitude)	4°32'S 15°30'E
Currency	zaire
Population (1995 est., in thousands)	43814
Surface area (square kms)	2344858
Population density (pop. per square km)	16
Sex ratio (females per 100 males)	102
Largest city (pop. in thousands)	Kinshasa (3455)
UN membership date	20-Sep-1960
Major language(s)	French, Luba, Congo, Mongo-Nkundu, Rwanda-Rundi, Lingala

Economic indicators	1985	1994
Exchange rate (US$)	56.00	35.00[a]
Balance of payments, current account (million US$)	−289	−643[b]
Tourist arrivals (in thousands)	39[c]	22[d]
GDP (million US$)	2953	3594[e]
GDP (per capita US$)	93	93[e]
Long-term rate of change in GDP (% per annum)	2.6	−2.8[e]
Economically active female population (%)	48.1[f]	43.5
Economically active male population (%)	86.9[f]	84.2
Annual growth of econ. active female pop. (%)	1.4[b]	...
Annual growth of econ. active male pop. (%)	2.8[b]	...
Agricultural production index (1979-1981=100)	125[c]	134[d]
Food production index (1979-1981=100)	125[c]	134[d]
Commercial energy production (1,000 Mt coal equiv)	2120[f]	2826[e]
Motor vehicles (per 1,000 population)	5	...
Telephone lines (per 100 inhabitants)	0.1[g]	0.1[e]

Social indicators	1990/95
Growth rate of population (% per annum)	3.2
Age group 0-14 years (%)	48.1
Age group 60+ years (women and men, %)	2.6/2.0
Life expectancy at birth (women and men, years)	53/50
Infant mortality rate (per 1,000 births)	93
Total fertility rate (births per woman)	7
Urban population (%)	29
Urban population growth rate (% per annum)	3.9
Rural population growth rate (% per annum)	2.9
Foreign-born (1985,%)	4.2
Refugees	391100
Government education expenditure (% of GDP)	1[c]
Primary-secondary gross enrolment ratio (f and m, per 100)	44/64[g]
Third-level students (per 100,000 population)	184[c]
Newspaper circulation (per 1,000 population)	2[f]
Television receivers (per 1,000 population)	1
Parliamentary seats (women and men, %)	5/95[g]

Environment	1990/91
Threatened species	82
Forested area (%)	74.3
CO_2 emissions (10,000 Mt)	1156
Energy consumption (1,000 Mt coal equiv)	65
Precipitation (mm)	1371[g]

a 1993. b 1990. c 1988. d 1992. e 1991. f 1980. g 1987.

Zambia

Region	Eastern Africa
Location (longitude, latitude)	15°41'S 28°30'E
Currency	Zambian kwacha
Population (1995 est., in thousands)	9381
Surface area (square kms)	752618
Population density (pop. per square km)	12
Sex ratio (females per 100 males)	102
Largest city (pop. in thousands)	Lusaka (979)
UN membership date	01-Dec-1964
Major language(s)	English, Bemba, Tonga, Nyanja, Lozi

Economic indicators	1985	1994
Exchange rate (US$)	5.70	500.00[a]
Consumer price index (1980=100)	253	38929[b]
Industrial production index (1980 = 100)	99	98[c]
Balance of payments, current account (million US$)	−398	−307[d]
Tourist arrivals (in thousands)	108[e]	159[c]
GDP (million US$)	3463	3799[d]
GDP (per capita US$)	505	453[d]
Long-term rate of change in GDP (% per annum)	0.0	−2.0[d]
Gross fixed capital formation (% of GDP)	15.5	24.6[d]
Economically active female population (%)	29.7[f]	...
Economically active male population (%)	88.5[f]	...
Annual growth of econ. active female pop. (%)	3.5[g]	...
Annual growth of econ. active male pop. (%)	3.0[g]	...
Labour force in industry (%)	42.9[f]	37.4[h]
Labour force in agriculture (%)	8.6[f]	10.3[h]
Agricultural production index (1979-1981=100)	142[e]	117[c]
Food production index (1979-1981=100)	139[e]	117[c]
Commercial energy production (1,000 Mt coal equiv)	1606[f]	1270[d]
Motor vehicles (per 1,000 population)	16	...
Telephone lines (per 100 inhabitants)	0.7[i]	0.8[d]

Social indicators	1990/95
Growth rate of population (% per annum)	2.8
Age group 0-14 years (%)	48.5
Age group 60+ years (women and men, %)	1.8/1.8
Life expectancy at birth (women and men, years)	45/43
Infant mortality rate (per 1,000 births)	84
Total fertility rate (births per woman)	6
Contraceptive use (% of currently married women)	15
Urban population (%)	43
Urban population growth rate (% per annum)	3.3
Rural population growth rate (% per annum)	2.5
Foreign-born (1985,%)	4.5
Refugees	142100
Government education expenditure (% of GDP)	3
Primary-secondary gross enrolment ratio (f and m, per 100)	62/72[e]
Third-level students (per 100,000 population)	178[h]
Newspaper circulation (per 1,000 population)	12
Television receivers (per 1,000 population)	26
Intentional homicides (1986, per 100,000 population)	10
Parliamentary seats (women and men, %)	7/93

Environment	1990/91
Threatened species	18
Forested area (%)	38.3
CO_2 emissions (10,000 Mt)	663
Energy consumption (1,000 Mt coal equiv)	205
Precipitation (mm)	923[j]
Average temperature (January and July, centigrade)	20.9/17.8[f]

a 1993. b February 1994. c 1992. d 1991. e 1988. f 1980. g 1990. h 1989. i 1987. j Broken Hill.

Zimbabwe

Region	Eastern Africa
Location (longitude, latitude)	17°56'S 31°06'E
Population (1995 est., in thousands)	11536
Sex ratio (females per 100 males)	102
Largest city (pop. in thousands)	Harare (855)
Major language(s)	English, Shona, Ndebele

Economic indicators	1985	1994
Exchange rate (US$)	1.64	8.32
Consumer price index (1980=100)	201	368[a]
Industrial production index (1980 = 100)	108	106[b]
Balance of payments, current account (million US$)	−76	−489[c]
Tourist arrivals (in thousands)	449[d]	738[e]
GDP (million US$)	4525	6194[c]
GDP (per capita US$)	541	603[c]
Long-term rate of change in GDP (% per annum)	6.9	1.9[c]
Economically active female population (%)	47.9[f]	42.6
Economically active male population (%)	88.4[f]	87.3
Annual growth of econ. active female pop. (%)	2.5[a]	...
Annual growth of econ. active male pop. (%)	3.2[a]	...
Labour force in industry (%)	27.2[f]	28.7[e]
Labour force in agriculture (%)	32.4[f]	23.9[e]
Agricultural production index (1979-1981=100)	139[d]	81[e]
Food production index (1979-1981=100)	133[d]	60[e]
Commercial energy production (1,000 Mt coal equiv)	3626[f]	6033[c]
Motor vehicles (per 1,000 population)	38	39[c]
Telephone lines (per 100 inhabitants)	1.3[f]	1.2[c]

Social indicators	1990/95
Growth rate of population (% per annum)	3.0
Age group 0-14 years (%)	44.6
Age group 60+ years (women and men, %)	2.3/2.0
Life expectancy at birth (women and men, years)	57/54
Infant mortality rate (per 1,000 births)	59
Total fertility rate (births per woman)	5
Contraceptive use (% of currently married women)	43[h]
Urban population (%)	32
Urban population growth rate (% per annum)	5.3
Rural population growth rate (% per annum)	1.9
Foreign-born (1985,%)	6.1
Refugees	137200
Government education expenditure (% of GDP)	8[h]
Primary-secondary gross enrolment ratio (f and m, per 100)	86/91
Third-level students (per 100,000 population)	585[h]
Newspaper circulation (per 1,000 population)	21
Television receivers (per 1,000 population)	26
Intentional homicides (1986, per 100,000 population)	12
Parliamentary seats (women and men, %)	12/88

Environment	1990/91
Threatened species	18
Forested area (%)	49.0
CO_2 emissions (10,000 Mt)	4635
Energy consumption (1,000 Mt coal equiv)	674
Precipitation (mm)	863[g]
Average temperature (January and July, centigrade)	20.0/13.6[f]

a 1990. b April 1994. c 1991. d 1988. e 1992. f 1980. g 1987. h 1989.

Technical notes

Geographical coverage

The designations employed and presentation of the material in this publication were adopted solely for the purpose of providing a convenient geographical basis for the accompanying statistical series.

The same qualification applies to all notes and explanations concerning the geographical units for which data are presented.

Because of space limitations, the country and area names listed in the tables are generally the commonly employed short titles in use in the United Nations, the full titles being used only when a short form is not available. Countries or areas are listed in alphabetical order.

Data relating to the People's Republic of China generally include those for Taiwan Province in the fields of statistics relating to population, area, natural resources and natural conditions such as climate. In other fields of statistics, they do not include Taiwan Province unless specifically stated.

Through accession of the German Democratic Republic to the Federal Republic of Germany with effect from 3 October 1990, the two German states have united to form one sovereign State. As from the date of unification, the Federal Republic of Germany acts in the United Nations under the designation "Germany".

On 22 May 1990 Democratic Yemen and Yemen merged to form a single State. Since that date they have been represented as one Member of the United Nations with the name "Yemen". Except as noted, data refer to the new State.

Data provided for Yugoslavia prior to 1 January 1992 refer to the Socialist Federal Republic of Yugoslavia, which was composed of six republics. Data provided for Yugoslavia after that date refer to the Federal Republic of Yugoslavia, which is composed of two republics (Serbia and Montenegro).

In 1991, the Union of Soviet Socialist Republics formally dissolved into fifteen independent countries (Armenia, Azerbaijan, Belarus, Estonia, Georgia, Kazakhstan, Kyrgyzstan, Latvia, Lithuania, Republic of Moldova, Russian Federation, Tajikistan, Turkmenistan, Ukraine and Uzbekistan). Whenever possible, data are shown for the individual countries. Otherwise, data for the former USSR are shown under the Russian Federation, with a footnote.

General indicators

Region presents seven regional groupings of countries and areas based mainly on continents. This series is from Annex I in the United Nations Statistical Yearbook. [14]*

Location provides the longitude and latitude of the largest city or agglomeration. This series is from the publication entitled *Climatological Normals (CLINO) for Climat and Climat Ship Stations for the Period 1931-1960.* [20]

Currency shows the national monetary unit. This information is from table 52 in the United Nations *Monthly Bulletin of Statistics.* [12]

Population estimates were prepared by the Population Division of the United Nations Secretariat and published in *World Population Prospects 1992.* [15]

Surface area represents the total surface area, comprising land area and inland waters (assumed to consist of major rivers and lakes), and excluding polar regions and uninhabited islands. This series is from table 3 in the United Nations *Demographic Yearbook.* [7]

Population density refers to population per square kilometre of the surface area. This series is from table A.19 in *World Population Prospects 1992.* [15]

Sex ratio is the number of females per 100 males. This series was calculated from data prepared by the Population Division of the United Nations Secretariat and published in table A.19 of *World Population Prospects 1992.* [15]

Largest city or agglomeration shows data for the largest city according to its administrative boundaries, or, if unavailable, largest urban agglomeration (city plus contiguous built-up areas) for each country or area. This series is from table A.11 in *World Urbanization Prospects 1992.* [16]

UN membership date is from the United Nations publication *Basic Facts About the United Nations,* 1993. [5]

Major languages are from *Countries, Peoples and Their Languages,* 1992. [2]

Economic indicators

Exchange rate is the amount of one currency required to purchase a fixed amount of another currency, as compiled by the International Monetary Fund and published as table 52 in the *Monthly Bulletin of Statistics.* [12] Data are generally shown in units of national currency

---

* Numbers in brackets refer to numbered entries listed in "Statistical sources" at the end of this publication.

per one US dollar and refer to the end-of-period quotation. For most countries, mid-point rates (i.e. the average of buying and selling rates) are shown.

Consumer price index numbers as published in table 51 in the *Monthly Bulletin of Statistics* [12] are designed to show changes over time in the cost of selected goods and services that are considered as representative of the consumption habits of the population concerned. The indices generally refer to "all items" and to the country as a whole.

The *industrial production index* generally covers mining, manufacturing and electricity, gas and water, and does not include construction unless otherwise indicated. This series is from table 26 in the United Nations *Statistical Yearbook*. [14]

Unemployment is defined to include persons above a certain age who during a specified period of time were without work, currently available for work and seeking work. For various reasons, national definitions of unemployment often differ from the recommended international standard definitions and thereby limit international comparability. Intercountry comparisons are also complicated by the variety of types of data collection systems used to obtain information on unemployed persons. These data are compiled by the International Labour Office and published as table 8 in the *Monthly Bulletin of Statistics*. [12]

Balance of payments, current account: This series refers to the current account balance and shows, for each country, the merchandise trade balance plus the balance on services, income and private unrequited and official unrequited transfers not included elsewhere. This series is from the International Monetary Fund's *International Financial Statistics*. [4]

Tourist arrivals data are those compiled by the World Tourism Organization and published in the United Nations *Statistical Yearbook*. [14] Tourists are defined as persons travelling for pleasure, domestic reasons, health, meetings, business, study, and the like, and stopping for period of 24 hours or more but less than one year in a country or area other than that in which they usually reside. They do not, therefore, include migrants, residents in a frontier zone, persons domiciled in one country or area and working in an adjoining country or area, travellers passing through a country or area without stopping overnight, transport crews and troops.

The estimates of total and per capita gross domestic product (*GDP*) in purchasers' values are expressed in

United States dollars. This series is from the national accounts database of the Statistical Division of the United Nations Secretariat. These estimates should be considered as measures of the total and per capita production of goods and services of the countries represented in economic terms and not as measures of the standard of living of their inhabitants.

GDP in purchasers' values is defined in the United Nations System of National Accounts as (a) the difference between the gross output value of resident producers minus the value of their intermediate consumption plus import duties, (b) the sum of domestic factor incomes (i.e. compensation of employees and operating surplus), indirect taxes net of subsidies and consumption of fixed capital, (c) the sum of government and private final consumption expenditure, gross capital formation in fixed assets and stocks and exports minus imports of goods and services.

Long-term rates of change in GDP are computed as average annual geometric rates of growth expressed in percentage form. They are based on the estimates of GDP at constant prices. This series is from the national accounts database of the Statistical Division of the United Nations Secretariat.

Gross fixed capital formation data are based on the percentage distribution of GDP in current prices. This series is from the national accounts database of the Statistical Division of the United Nations Secretariat.

The *economically active population* and *annual rate of change for economically active population* refer to the total of employed persons (including employers, persons working on their own account, salaried employees and wage earners and, in so far as data are available, unpaid family workers) and of unemployed persons at the time of the census or survey which provided the data. In general, the economically active population does not include full-time students who are not working, persons occupied solely in household work, retired persons living entirely on their own means and persons wholly dependent upon others. Percentages generally refer to males and females respectively, but if distribution by sex is not available, data refer to totals only. These series are from the draft estimates and projections by the International Labour Office (to be issued in final form in 1995), supplemented by data from the International Labour Office, *Year Book of Labour Statistics* [3] and national reports.

Labour force in industry data refer to the economically active population in industry expressed as a percentage of the total labour force.

Labour force in agriculture refers to population economically active in agriculture expressed as a percentage of the total labour force. These series are from table 3 in *Year Book of Labour Statistics* [3]. The comparability of labour force data is hampered by differences between countries. Not only are there different details in definitions used and groups covered, but also differences in methods of collection, classification and tabulation of data. For some countries, statistics cannot be compared over time since they are drawn from a variety of sources (International Labour Organization estimates, census data, survey data, national estimates)—all of which differ as to scope, coverage, reference period, time of year and so on.

The *agricultural production index* covers all crops and livestock products originating in each country on which information is available. This series is from table 5 in the Food and Agriculture Organization of the United Nations publication *FAO yearbook: Production.* [1]

The *food production index* covers commodities that are considered edible and contain nutrients. Coffee and tea are excluded because they have practically no nutritional value. The index numbers shown may differ from those produced by countries themselves because of differences in concepts of production, coverage, weights, time reference of data, and methods of evaluation. This series is from the Food and Agriculture Organization of the United Nations publication *FAO yearbook: Production.* [1]

Commercial energy production refers to the production of various forms of primary energy which are converted into a common unit (metric ton of coal equivalent). This series is from the energy statistics database of the Statistical Division of the United Nations Secretariat.

The *motor vehicles in use* data are from table 76 in the United Nations *Statistical Yearbook.* [14] This series refers to passenger cars and commercial vehicles in use according to census or registration figures for years census or annual registration took place. Passenger vehicles are vehicles seating no more than nine persons (including the driver), including taxis, jeeps and station wagons. Commercial vehicles include vans, lorries (trucks), buses, tractors and semi-trailer combinations.

Telephone lines refers to the number of main lines per 100 inhabitants. Main lines refer to telephone lines connecting a customer's equipment to the PSTN and which have a dedicated port on a telephone exchange. Main telephone lines per 100 inhabitants is calculated by dividing the number of main lines by the population and multiplying by 100. This series is from table 23 in the United Nations *Statistical Yearbook.* [14]

Largest export industries are the major exporting industries, expressed in percentages. This series is from table 2 in the United Nations *International Trade Statistics Yearbook.* [10]

Major export and import trading partners are expressed as a percentage of the total exports and imports of the country or area. These series are from table 3 in the United Nations *International Trade Statistics Yearbook.* [10]

Social indicators

The annual *growth rate of population* is the average annual percentage change in total population size in the designated period. These series are from table A.2 in *World Population Prospects 1992.* [15]

Age group 0-14 years refers to the population aged 0-14 years of both sexes as a percentage of total population. *Age group 60+ years* refers to elderly men as a percentage of all males and elderly women as a percentage of all females. These series are from table A.19 in the United Nations publication *World Population Prospects 1992.* [15] Supplementary data was obtained from the United Nations *Demographic Yearbook* [7] and footnoted DYB90.

Life expectancy at birth and *infant mortality rate* are from tables A.15 and A.16 respectively in *World Population Prospects 1992.* [15] The infant mortality rate is the number of infants who die before reaching 1 year of age, per thousand live births in a given year or period. The life expectancy at birth indicates the average number of years expected to be lived by a newborn baby, given the age-specific mortality rate in the year of birth.

Total fertility rate is the average number of children that would be born alive to a hypothetical cohort of women if, throughout their reproductive years, the age-specific fertility rates for the specified year remained unchanged. This series is from table A.13 in the United Nations publication *World Population Prospects 1992.* [15]

Contraceptive use refers to use by currently married women of any method, expressed as a percentage. The sources of data are the contraceptive use database compiled by the Population Division of the United Nations Secretariat and published in *Levels and Trends of Contraceptive Use as Assessed in 1988.* [11] Supplementary data were obtained from national health reports and surveys and footnoted as NATHR.

Urban population, urban population growth rate and *rural population growth rate* series are based on the number of persons defined as "urban" or "rural" according to national definitions of this concept. In most cases these definitions are those used in the most recent population census. These series are from tables A.1, A.5 and A.6 in *World Urbanization Prospects.* [16]

Foreign-born population refers to persons born outside the country or area in which they are enumerated. The country or area of birth is based on the national boundaries existing at the time of census. This series is from *Trends in the Total Migrant Stock*, a diskette database maintained by the Population Division of the United Nations Secretariat.

The term *refugees* refers to any person who, owing to well-founded fear of being persecuted for reasons of race, religion, nationality, membership of a particular social group or political opinion, is outside the country of his nationality and is unable or, owing to such fear, unwilling to avail himself of the protection of that country; or who, not having a nationality and being outside the country of his former habitual residence is unable or, owing to such fear, unwilling to return to it. This series is from the *The State of the World's Refugees: The Challenge of Protection.* [18]

Government education expenditures: This series is from the UNESCO *Statistical Yearbook* [17] and shows the general trends in public expenditure on public and private education expressed as a percentage of the gross national product. The data shown should be considered as approximate indications of the public resources allocated to education. These data are frequently revised and may differ from those shown in previous editions of *World Statistics in Brief.*

Primary-secondary gross enrolment ratio and third-level students: These series are from the UNESCO *Statistical Yearbook.* [17] For the first and second levels, the enrolment ratio generally is the total enrolment of all ages in first- and second-level education, divided by the total population in the official ages of enrolment in the country times 100. In some cases, the gross enrolment ratio will exceed 100 to the extent that the actual age distribution of pupils is wider than the official school ages.

Newspaper circulation refers to daily general interest newspapers, defined as publications devoted primarily to recording general news. They are considered to be "daily" if they appear at least four times a week. This series is compiled by UNESCO and also published in the United Nations *Statistical Yearbook*. [14]

Television receivers refers to television receivers in use and/or licenses issued per thousand inhabitants. This series is compiled by UNESCO and also published in the United Nations Statistical Yearbook. [14]

Intentional homicides refers to death purposely inflicted by another person per 100,000 population. Data was compiled by the Crime Prevention and Criminal Justice Branch, Centre for Social Development and Humanitarian Affairs of the United Nations Secretariat, from national responses to questionnaires in the First, Second and Third United Nations Surveys of Crime Trends, Operations of Criminal Justice Systems and Crime Prevention Strategies. This series was published as table 34 in the United Nations *Compendium of Social Statistics* [6] Supplementary data footnoted as DYB92 was obtained from the United Nations *Demographic Yearbook*. [7] where *homicide and injury purposely inflicted by other persons* is reported as a cause of death.

Parliamentary seats refers to the number of women and men in the lower chamber of parliament, expressed as a percentage. Data was obtained from the Inter-Parliamentary Union, Distribution of seats between men and women in National Parliaments, Statistical data from 1945 to 30 June 1991, Series Reports and Documents, no. 18 (Geneva, 1991), Distribution of seats between men and women in the 144 national assemblies, Series Reports and Documents, no. 14 (Geneva, 1987) and Women in Parliament as of 30 June 1993 (World Map) (Geneva, 1993).

Environmental indicators

Threatened species includes those that are listed as endangered, vulnerable, rare, indeterminate or insufficiently known. Data on the number of threatened species are derived from the 1994 International Union for Conservation of Native and Natural Resources *Red List for Animals*. [19] This series was obtained from the

United Nations Environment Programme/GEMS Monitoring and Assessment Research Centre, *Environment Data Report 1993/94* and published in the United Nations *Statistical Yearbook.* [14]

Forested area refers to land under natural or planted stands of trees, whether or not productive. This series was obtained from the Food and Agriculture Organization of the United Nations (Rome) and published in the United Nations *Statistical Yearbook.* [14]

CO_2 emissions from fossil fuel combustion and cement manufacturing are derived from UN consumption data for gas, liquid and solid fuels plus cement manufacturing and gas flaring statistics to which appropriate emission factors have been applied. This series was obtained from the United Nations Environment Programme/GEMS Monitoring and Assessment Research Centre, *Environment Data Report 1993/94.* and published in the United Nations *Statistical Yearbook.* [14]

Energy consumption data refer to "apparent consumption" and are derived from the formula "production + imports − exports − bunkers +/− stock changes". Accordingly, the series on apparent consumption may in some cases represent only an indication of the magnitude of actual inland availability. This series was obtained from the energy statistics database of the Statistical Division of the United Nations Secretariat.

Precipitation and *average temperature* are measurements from the weather stations closest to the largest city or agglomeration. These series are from the publication entitled *Climatological Normals (CLINO) for Climat and Climat Ship Stations for the Period 1931-1960* issued in 1971 and reprinted in 1982 by the World Meteorological Organization. [20] This publication was prepared to provide climatological normals computed from the period 1931-1960 in order that national meteorological services could have some average or normal value to compare with current climatological information.

Statistical sources

[1] Food and Agriculture Organization of the United Nations, *FAO yearbook: Production 1992* (Rome).

[2] Gunnemark, E. V., *Countries, Peoples and Their Languages*, revised edition (Gothenberg, 1992).

[3] International Labour Office, *Year Book of Labour Statistics* (Geneva, various years up to *1993*).

[4] International Monetary Fund, *International Financial Statistics* (monthly) (Washington, DC, September 1994).

[5] United Nations, *Basic Facts about the United Nations*, 1993.

[6] ——, *Compendium of Social Statistics and Indicators 1988* (Series K, United Nations publication, 1991).

[7] ——, *Demographic Yearbook* (Series R, United Nations publication, various years up to *1992*).

[8] ——, *Energy Statistics Yearbook* (Series J, United Nations publication, various years up to *1991*).

[9] ——, *Industrial Statistics Yearbook* (Series P, United Nations publication, various years up to *1991*).

[10] ——, *International Trade Statistics Yearbook* (Series G, United Nations publication, various years up to *1992*).

[11] ——, *Levels and Trends of Contraceptive Use as Assessed in 1988* (United Nations publication, Sales No. E.89.XIII.4).

[12] ——, *Monthly Bulletin of Statistics* (Series Q, United Nations publication, various issues up to December 1994).

[13] ——, *National Accounts Statistics: Analysis of Main Aggregates* (Series X, United Nations publication, various years up to *1991*).

[14] ——, *Statistical Yearbook 1992* (Series S, United Nations publication, 39th edition).

[15] ——, *World Population Prospects: The 1992 Revision* (United Nations publication, Sales No. E.93.XIII.7).

[16] ——, *World Urbanization Prospects: The 1992 Revision* (United Nations publication, Sales No. E.93.XIII.11).

[17] United Nations Educational, Scientific and Cultural Organization, *Statistical Yearbook* (Paris, various years up to *1993*).

[18] United Nations High Commissioner for Refugees, *The State of the World's Refugees: The Challenge of Protection* (New York and London, 1993).

[19] World Conservation Union, *1994 ICUN Red List of Threatened Animals* (Gland and Cambridge, 1994).

[20] World Meteorological Organization, *Climatological Normals (CLINO) for Climat and Climat Ship Stations for the Period 1931-1960* (Geneva,1982).

[21] World Tourism Organization, *Yearbook of Tourism Statistics 1993* (Madrid).

Current United Nations statistical publications

Monthly Bulletin of Statistics
Series Q, Volume XLIX, 1995 (E/F). Annual subscription: $450.00; $60.00 per copy
Provides monthly statistics on 60 subjects from over 200 countries and territories, together with special tables illustrating important economic developments. Quarterly data for significant world and regional aggregates are included regularly.

Statistical Yearbook, thirty-ninth issue
Series S, No. 15, 1055 pages, $110.00, Sales No. E/F.94.XVII.1
This annual compilation of statistics for all countries and areas of the world provides data on world and regions; population and social statistics; economic activity; and international economic relations. This issue presents 104 tables in the fields of demographic and social statistics, national accounts, finance, labour force, wages and prices, agriculture, industry, transport and communications, energy, environment, international merchandise trade and tourism, and covers, in general, the ten-year period 1982-1991 or 1983-1992, with data available as of 31 December 1993.

Statistical Yearbook, thirty-ninth issue, on CD-ROM
(SYB-CD) $199.00, Sales No. E.95.XVII.5
SYB-CD contains the same information as the *Statistical Yearbook*, thirty-ninth issue, but is in a database format on CD-ROM. Resident software provides easy data extraction and provides a variety of tabulation formats, graphing, export and printing functions.

StatBase Locator on Disk: UNSTAT's Guide to International Computerized Statistical Databases
World Statistics in Brief — Special Issue (V/15)
$20.00, Sales No. E.94.XVII.8
StatBase Locator on Disk provides an inventory of international statistical databases from which data are available in electronic formats. It is contained on one 3 1/2" diskette for IBM-compatible microcomputers and provides information on the organization, content, producing agency and means of access for each database. All fields of social and economic statistics with internationally available data are covered for statistical databases of all United Nations agencies, Eurostat (European Union) and OECD. *StatBase Locator on Disk* is self-contained, with user-friendly software for searching and retrieving database information.

Demographic Yearbook 1992
Series R, No. 23, 823 pages, $125.00, Sales No. E/F.94.XIII.1
Contains comprehensive demographic and related statistics for over 200 countries or areas of the world. This issue features fertility and mortality, with tables on induced abortion; foetal, infant and maternal mortality; deaths by age, sex, residence, cause of death; and life expectancy; and on births, crude birth rates, births by age of mother, sex and residence and age-specific fertility rates; and the latest statistics on population size, structure and marriage and divorce.

Demographic Yearbook Special Issue:
Population Ageing and the Situation of Elderly Persons

Series R, No. 22, 855 pages, $75.00, Sales No. E/F.92.XIII.9

This special issue of the *Demographic Yearbook* focuses on population ageing and elderly persons, using 40 years of national population statistics on population age structure and residence, population ageing, living arrangements, socio-demographic characteristics and economic characteristics, and disability and mortality statistics.

It also includes articles on age structure changes, mortality trends among the elderly, and disability statistics and household and family statistics in ageing.

Population and Vital Statistics Report

Series A, Vol. XLVII, 1995 (E) quarterly. Annual subscription: $30.00; $10.00 per copy.

This publication presents in one table series on total population and births, deaths and infant deaths for over 200 countries or areas.

National Accounts Statistics:
Main Aggregates and Detailed Tables, 1991

Series X, No. 20, Parts I and II (not sold separately), 2,122 pages total, $125.00. Thirty-fifth issue, Sales No. E.94.XVII.5

Contains detailed national accounts estimates for 178 countries and areas. The estimates for each country and area are presented with uniform table headings and classifications as recommended in the United Nations System of National Accounts (SNA).

Trends in International Distribution of Gross World Product

National Accounts Statistics—Special Issue

Series X, No. 18, 336 pages, $50.00, Sales No. E.92.XVII.7.

This publication analyses trends in international distribution of gross world product during the last two decades. It focuses on how trends differ when applying conversion rates other than exchange rates. It deals with economic performance of countries or areas and regions and inequality in GDP distribution among countries and regions between 1970 and 1989. Detailed country tables are included.

1992 International Trade Statistics Yearbook

Clothbound, $135.00, Volumes I and II (not sold separately)

Volume I. Trade by Country

Series G, No. 41, 1130 pages, Sales No. E/F.94.XVII.3, Vol. I

The forty-first issue provides basic information on countries' external trade performance and trends in current value and volume and price, trading partners and significant commodities imported or exported. Basic summary tables show the contribution of trade of each country to the trade of its region and of the world. Contains data for 153 countries or customs areas.

Volume II. Trade by Commodity

Series G, No. 41, Add.1, 804 pages, Sales No. E/F.94.XVII.3, Vol. II

Volume II contains commodity tables showing the total economic world trade of certain commodities, analysed by regions and countries.

Commodity Trade Statistics

Series D. Annual subscription: $470.00; $30.00 per fascicle (E)

Issued in fascicles of about 250 pages as annual data become available in the 625 sub-groups of the United Nations Standard International Trade Classification. Approximately 28 fascicles annually. Quarterly and annual data are also available on microfiches from the United Nations Statistical Division, New York.

Energy Statistics Yearbook, 1992

Series J, No. 36, 546 pages, $85.00, Sales No. E/F.94.XVII.9

This thirty-sixth issue of internationally comparable series on commercial energy summarizes world energy trends for approximately

215 countries and areas. It includes energy data for States of the former Soviet Union in Europe and Asia. It contains data in original and common units (coal equivalent, oil equivalent, joules) for the years 1989-1992. Annual data are presented on the production, trade and consumption of solid, liquid and gaseous fuels and electricity. Per capita consumption series are included for all products.

Energy Balances and Electricity Profiles 1992

Series W, No. 7, 520 pages, paperbound $75.00, Sales No. E/F.94.XVII.14

This is the seventh issue of energy data for developing countries in the format of overall energy balances and electricity profiles. Included are energy balances for 49 countries or areas and special electricity profiles for 75 countries or areas.

Industrial Commodity Statistics Yearbook, 1992

Series P, No. 32, 1047 pages, $110.00, Sales No. E/F.94.XVII.13

Provides statistics on the production of about 530 industrial commodities for approximately 200 countries or areas and the regional and grand totals for most of the statistical series. Also contains data by country on apparent consumption of about 230 industrial commodities for 1988-1992. The index of commodities, an inventory of available data by country and a table of correspondence between ISIC-based, SITC and Harmonized Commodity Description and Coding System (HS) codes are shown in the annexes.

United Nations Women's Indicators and Statistics Database (Wistat), Version 3

Sales No. E.95.XVII.6, $149 for diskettes or CD-ROM

Wistat contains detailed national statistics on a wide range of social and economic topics, disaggregated by sex, for 220 countries or areas of the world. It includes information available at the international level for the period 1970-1994, with some projections to 2025. The complete database is available on diskettes (in spreadsheets) or CD-ROM. The CD-ROM runs with resident software for easy data extraction and provides a variety of tabulation formats and graphing, export and printing capabilities. *Wistat-CD* may also be installed on a local area network (LAN) at no additional cost.

The World's Women 1970-1990: Trends and Statistics

Series K, No. 8, 135 pages, $19.95, Sales No. 90.XVII.3
(A, C, E, F, R, S)

This unique publication highlights indicators on women's conditions worldwide, in a form that non-specialists can readily understand. It provides numbers and analyses to inform people everywhere about how much women contribute to economic life, political life and family life—and it provides the ammunition to persuade public and private decision makers to change policies that are unfair to women.